Lecture Notes in Energy

Volume 1

Lecture Notes in Energy (LNE) is a series that reports on new developments in the study of energy: from science and engineering to the analysis of energy policy. The series' scope includes but is not limited to, renewable and green energy, nuclear, fossil fuels and carbon capture, energy systems, energy storage and harvesting, batteries and fuel cells, power systems, energy efficiency, energy in buildings, energy policy, as well as energy-related topics in economics, management and transportation. Books published in LNE are original and timely and bridge between advanced textbooks and the forefront of research. Readers of LNE include postgraduate students and non-specialist researchers wishing to gain an accessible introduction to a field of research as well as professionals and researchers with a need for an up-to-date reference book on a well-defined topic. The series publishes single- and multi-authored volumes as well as advanced textbooks.

Robert Basmadjian

Demand-Side Energy and Power Management in Data Centers

Are Data Centers Modern Societies' Double-Edged Sword?

Springer

Robert Basmadjian
College of Computing
Mohammed VI Polytechnic University
(UM6P)
Benguerir, Morocco

ISSN 2195-1284 ISSN 2195-1292 (electronic)
Lecture Notes in Energy
ISBN 978-3-031-85955-7 ISBN 978-3-031-85956-4 (eBook)
https://doi.org/10.1007/978-3-031-85956-4

This Springer imprint is published by the registered company Springer Nature Switzerland AG
The registered company address is: Gewerbestrasse 11, 6330 Cham, Switzerland

Preface

In an era defined by an insatiable appetite for data processing and communication, our reliance on data centers as the backbone of digital infrastructure cannot be overstated. As organizations worldwide increasingly shift towards cloud/high-performance computing and data-driven decision-making, the challenge of energy and power management in these facilities has become a focal point of both operational efficiency and sustainability. This book is a synthesis of my 15 years of experience in this vital field, aimed at providing a thorough understanding of energy management through both theoretical frameworks and practical applications.

The structure of the book is thoughtfully designed to guide you through a journey of discovery and insight. In the first part, we will establish a foundation by exploring the core principles of energy systems (e.g., Smart Grid) as well as demand-side management. We will trace the evolution of our power system from traditional models to the Smart Grid, emphasizing the critical shift from a "supply follows demand" approach to one that prioritizes demand-side management. This part will also equip readers with essential concepts and introduce various pricing and incentive-based schemes that drive effective demand-side management.

The second part takes a deeper dive into demand-side management and flexibility opportunities specifically in data centers. This section is not merely an academic exercise; it draws on real-world scenarios that I encountered during my years of involvement in European Union R&D projects. Thus, I will share case studies that illustrate the successful implementation of innovative solutions such as energy management systems (EMS). These examples are drawn from my practical experiences and collaborations with industry leaders, showcasing how theoretical knowledge can be transformed into tangible results.

In this book's third part, we will turn our gaze towards methodologies for predicting server power demand and analyzing the power-performance trade-off. We'll begin by providing essential background on CMOS circuits, the fundamental building blocks of our computing and communication technologies. This includes a comprehensive discussion on estimating power demand in CMOS circuits and a detailed examination of power demand models for servers, covering components such as processors, memory, and storage, among others. Finally, we will introduce analytical methods, including Markovian analysis and Markov chains, to model and evaluate the power-performance optimizations of servers utilizing dynamic voltage and frequency scaling (DVFS) technology.

In the last part and concluding chapter of the book, we will shift our focus to the future. The landscape of data centers is rapidly evolving, influenced by technological advancements and societal demands for sustainability. We will discuss the trends that are shaping this evolution, such as the rise of edge computing, renewable energy sources, ARM-based technology, and AI-driven energy management systems. I will share my vision for the future, emphasizing the importance of adaptability and resilience in an industry that is both challenging and rewarding.

The insights shared in this book stem not only from my professional experiences—currently as the Head of the Toubkal Supercomputer at UM6P in Morocco—but are also shaped by my role as a lecturer at various esteemed institutions across France, Germany, Austria, and now the College of Computing at UM6P as an Associate Professor. Engaging with inquisitive students has been a profound source of inspiration, as their enthusiasm for confronting contemporary challenges has urged me to continuously refine my understanding of energy management.

Their diverse questions and unique perspectives have enriched my approach and deepened my commitment to this field. I aspire for this book to be a valuable resource for students, researchers, academicians, practitioners, and anyone passionate about the vital intersection of technology and sustainability, empowering them to navigate and innovate within this critical domain.

As we embark on this exploration of energy and power management in data centers, I invite you to join me in envisioning a future where our digital infrastructure operates with unparalleled efficiency and minimal environmental impact. Together, let us work towards solutions that not only meet the demands of today but also pave the way for a sustainable tomorrow.

Passau, Germany
December 2024

Robert Basmadjian

Acknowledgements I would like to dedicate this work, to my late father, Meguerditch. His steadfast support and belief in my dreams shaped who I am today. From our discussions about life, science, politics, and literature to his encouragement during my darkest moments, he instilled in me a love for storytelling that I carry with me always. This book is a testament to his enduring influence and the values he imparted. Thank you, Dad, for being my guiding light.

I want to express my heartfelt gratitude to everyone who has supported me on this incredible journey. To my amazing wife, Meghedi, your unwavering love and encouragement have been my greatest pillar of strength. To my sons, Alec, Marc, and Bernard, your boundless joy and curiosity inspire me every day, reminding me of the magic of storytelling that I hope to share with you for years to come. To my family—my mother, Rita, whose endless support and love have provided me with a solid foundation, and my brother, whose thoughtful insights have kept me grounded—I am eternally thankful. To my close ones—Avedis and Krikor Demirdjian, Georges Rammouz, Hikmat Achkar, Sevag Mateyan, Ari Derbedrossian, and Oleg Kaplanian—thank you for your unwavering belief in me and for being there whenever I needed a helping hand. Your friendship means the world to me. I would like to extend my gratitude to Oleg and my brother for their insightful feedback, particularly on the initial chapters of this book. Their contributions have been instrumental in refining my thoughts and ideas.

To my colleagues and fellow peers, especially within the EU-funded projects of FIT4Green and ALL4Green, your insights and camaraderie have enriched this project. In this respect, the dedication and professionalism of Marco Di Girolamo and Giovanni Giuliani have motivated me to push my limits. Lastly, I want to extend my sincere gratitude to my colleagues at the College of Computing and the Toubkal Supercomputing Center at the UM6P. Your unwavering support and collaboration have truly made my workdays more manageable and enjoyable. I appreciate the teamwork, dedication, and companionship we share, which significantly enhance our productivity and make tackling challenges a breeze. Thank you for creating such a positive and supportive environment!

I would like to express my heartfelt thanks to Karim Boubouh for his invaluable comments and support in creating the figures for this book. The illustrations were crafted using the open-source online drawing tool Excalidraw (https://excalidraw.com/) and the open-source icon generator Reshot (https://www.reshot.com/). Thank you both communities for making these tools available as open access.

Competing Interests The author has no competing interests to declare that are relevant to the content of this manuscript.

Contents

Acronyms

AC	Air Conditioner
ADC	Analog to Digital Converter
AI	Artificial Intelligence
ALU	Arithmetic Logic Unit
AMR	Automatic Meter Reading
API	Application Programming Interface
ARIMA	Auto-Regressive Integrated Moving Average
ARIMAX	Auto-Regressive Integrated Moving Average with eXogenous variables
ARM	Advanced RISC Machine
ASD	Adjustable Speed Drives
ASHRAE	American Society of Heating, Refrigerating and Air-Conditioning Engineers
ASM	Advanced Smart Meter
AWS	Amazon Web Services
B5G	Beyond 5th Generation
BESS	Battery Energy Storage System
BMS	Building Management System
CAES	Compressed Air Energy Storage
CAS	Column Address Strobe
CC	Cloud Computing
CDMA	Code Division Multiple Access
CHP	Combined Heat and Power
CL	Centralized Learning
CMOS	Complementary Metal-Oxide-Semiconductor
CNP	Contract Net Protocol
CO_2	Carbon Dioxide
COVID-19	Coronavirus Disease
CPP	Critical Peak Pricing
CPS	Cyber-Physical Systems
CPU	Central Processing Unit

CRAC	Computer Room Air Conditioner
CRAH	Computer Room Air Handler
CSMS	Charging Station Management System
CTMC	Continuous Time Markov Chain
CU	Control Unit
CUE	Carbon Usage Effectiveness
DC	Data Concentrator
DCO	Data Center Operator
DDR	Double Data Rate
DER	Distributed Energy Resources
DLC	Direct Load Control
DR	Demand Response
DSM	Demand-Side Management
DSO	Distribution System Operator
DTMC	Discrete Time Markov Chain
DVFS	Dynamic Voltage and Frequency Scaling
EC2U	Elastic Cloud Compute Unit
ECC	Error Correcting Codes
EDVAC	Electronic Discrete Variable Automatic Computer
EE	Energy Efficiency
EEE	Energy-Efficient Ethernet
EMS	Energy Management System
EMSA	Electric Mobility System Architecture
ENIAC	Electronic Numerical Integrator and Computer
ESS	Energy Storage System
EU	European Union
EV	Electric Vehicle
FCFS	First Come First Served
FinFET	Fin Field-Effect Transistor
FL	Federated Learning
G2V	Grid to Vehicle
GDPR	General Data Protection Regulation
GHI	Global Horizontal Irradiance
GMS	Gas Management System
GO	Grid Operator
GPRS	General Packet Radio Service
GPU	Graphics Processing Unit
GRU	Gated Recurrent Unit
GSM	Global System for Mobile Communications
GUI	Graphical User Interface
HAN	Home Area Network
HCPS	Human-centered Cyber-Physical Systems
HDD	Hard Disk Drive
HDMI	High-Definition Multimedia Interface
HEMS	Home Energy Management System

HHMR	Hand-Held Meter Reader
HMI	Human-Machine Interface
HMS	Heat Management System
HPC	High-Performance Computing
HPL	High Performance Linpack
HT	Hoch Tarif
HVAC	Heating Ventilation and Air Conditioning
ICT	Information and Communication Technologies
IEC	International Electrotechnical Commission
IoT	Internet of Things
IP	Internet Protocol
IT	Information Technology
ITC	IT Customer
JSON	JavaScript Object Notation
KPI	Key Performance Indicator
KVM	Kernel-based Virtual Machine
LCFS	Last Come First Served
LCS	Load Control Switch
LSTM	Long Short-Term Memory
LTE	Long-Term Evolution
MC	Markov Chain
MDMS	Meter Data Management System
ML	Machine Learning
MMU	Memory Management Unit
NAN	Neighborhood Area Network
NAS	Network Attached Storage
NIC	Network Interface Card
NP	Nondeterministic Polynomial
NT	Nieder Tarif
NVMe	Non-volatile Memory Express
OpenADR	Open Automated Demand Response
Opex	Operational Expenditure
OS	Operating System
P2P	Peer-to-Peer
PC	Personal Computer
PCIe	Peripheral Component Interconnect Express
PLC	Power Line Carrier
PoC	Proof of Concept
PSTN	Public Switched Telephone Network
PSU	Power Supply Unit
PUE	Power Usage Effectiveness
PV	Photovoltaic
QoS	Quality of Service
QSFP	Quad Small Form-factor Pluggable
RAM	Random Access Memory

RAS	Row Address Strobe
RDBMS	Relational Database Management Systems
RDMA	Remote Direct Memory Access
RES	Renewable Energy Sources
RF	Radio Frequency
RISC	Reduced Instruction Set Computing
RJ45	Registered Jack-45
RL	Reinforcement Learning
RNN	Recurrent Neural Network
RPM	Rotations Per Minute
RR	Round Robin
RTP	Real-Time Pricing
SAN	Storage Area Network
SARIMA	Seasonal Auto-Regressive Integrated Moving Average
SARIMAX	Seasonal Auto-Regressive Integrated Moving Average with eXogenous variables
SAS	Serial Attached SCSI
SATA	Serial Advanced Technology Attachment
SDA	Supply Demand Agreement
SET Plan	Strategic Energy Technology Plan
SFP+	Small Form-factor Pluggable
SG	Smart Grid
SGAM	Smart Grid Architecture Model
SLA	Service Level Agreement
SM	Smart Meter
SoC	State of Charge
SoCs	Systems-on-Chip Case
SoH	State of Health
SPoF	Single Point of Failure
SSD	Solid-State Drive
SSL	Secure Sockets Layer
ST	Smart Thermostat
TC	Traditional Computing
TCO	Total Cost of Ownership
TCP	Transmission Control Protocol
THD	Total Harmonic Distortion
ToU	Time of Use
TV	Television
UPS	Uninterruptible Power Supply
US	United States
USB	Universal Serial Bus
V2G	Vehicle to Grid
VM	Virtual Machine
VMM	Virtual Machine Monitor
WAN	Wide Area Network

WiMAX	World Interoperability for Microwave Access
XML	Extensible Markup Language
XSD	XML Schema Definition

Chapter 1
Introduction

1.1 Technological Evolution

Over the past four decades, we have witnessed remarkable technological advancements, largely propelled by the advent of the Internet in the 1980s. This pivotal development laid the groundwork for the emergence of distributed systems in the 1990s, which enabled the decentralization of computing resources and processes. By the early 2000s, the Internet of Things (IoT) had made its debut, effectively merging the advancements in both the Internet and distributed systems. This convergence made it possible to seamlessly incorporate small IoT devices—such as sensors that monitor light, temperature, humidity, and motion—into our everyday environments. These devices not only have the ability to collect raw data through their sensors but can also transmit this information externally. Moreover, they can receive data from outside sources and execute actions within their integrated systems using built-in actuators.

Around 2005, cloud computing emerged as a transformative computing paradigm, providing decentralized solutions for data storage (like Google Drive), computation (such as Google Colab), and communication. This innovation greatly enhanced the efficiency and accessibility of data management. Building upon this foundation, cloud computing soon extended its capabilities to the IoT realm through fog and edge computing. This allowed for data storage and processing to occur on powerful resources[1] located near IoT devices, significantly reducing latency and improving response times.

In the last ten years, the landscape of Big Data has evolved dramatically, driven by breakthroughs in machine learning (ML) and artificial intelligence (AI). These technologies have opened new frontiers, enabling us to extract valuable insights

[1] As we progress through this book, we will explore these formidable resources, which are referred to as data centers.

R. Basmadjian, *Demand-Side Energy and Power Management in Data Centers*,
Lecture Notes in Energy 1, https://doi.org/10.1007/978-3-031-85956-4_1

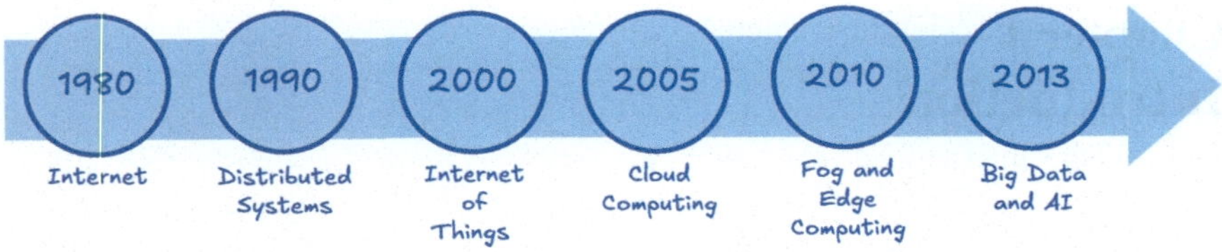

Fig. 1.1 The chronological evolution of various technologies

from vast amounts of data, thus enhancing decision-making processes across various sectors. The advancements in ML and AI have demonstrated their transformative potential, offering innovative solutions and efficiencies previously unattainable.

The chronological progression of these pivotal technological milestones is depicted in Fig. 1.1, highlighting the interconnected nature of each innovation and its role in shaping today's technological landscape. In this context, any single technology that underpins our everyday digital activities—ranging from computation and communication to data storage—is fundamentally reliant on data centers. For instance, in the modern business world, virtually every major corporation such as Google, Amazon, Meta, among others has established its own data center infrastructure to provide an array of services, ensuring they can meet the demands of their operations and customers. These data centers are crucial for processing vast amounts of data, hosting applications, and enabling seamless communication across various platforms.

From a scientific computing perspective, high-performance computing (HPC) data centers or so-called supercomputers represent significant investments made by governments and research institutions. These facilities are designed to showcase and harness immense computational power, serving as a testament to a nation's technological capabilities. They enable complex simulations, advanced data analysis, and innovative research across fields such as climate modeling, quantum computing, genomics, and artificial intelligence. By investing in these high-performance systems, countries not only enhance their scientific prowess but also foster collaboration across institutions and industries, driving progress and innovation on a global scale. The synergy between commercial data centers and high-performance computing facilities illustrates the foundational role of data center infrastructure in our increasingly data- and compute-driven world.

1.2 Human-Centered Cyber Physical System

The progress of information and communication technologies[2] (ICT) has played a crucial role in the emergence of cyber-physical systems (CPS). Essentially, these

---

[2] Examples of ICT resources encompass servers, storage devices, and networking equipment, all of which are typically housed within a data center.

systems comprise multiple interconnected sub-systems, each characterized by its complex nature, all intertwined through ICT. Moreover, human involvement in the control loop has resulted in the development of human-centered CPS (HCPS), highlighting the interdependence between humans and these systems, where human interventions have an impact on system behavior and vice versa.

A prominent example of an HCPS that we frequently encounter in this book is the Smart Grid (SG). In this context, the grid operator (GO)—a human being—assumes the responsibility of closely monitoring the system to prevent and minimize blackout events as much as possible. The Smart Grid itself comprises various subsystems, including generation sub-systems such as fossil-based and nuclear power plants, as well as renewable energy sources. Additionally, there are consumption subsystems like hospitals, industries, and data centers, to name a few. Notably, certain consumption sub-systems can also function as generation sub-systems, resulting in their classification as prosumers.

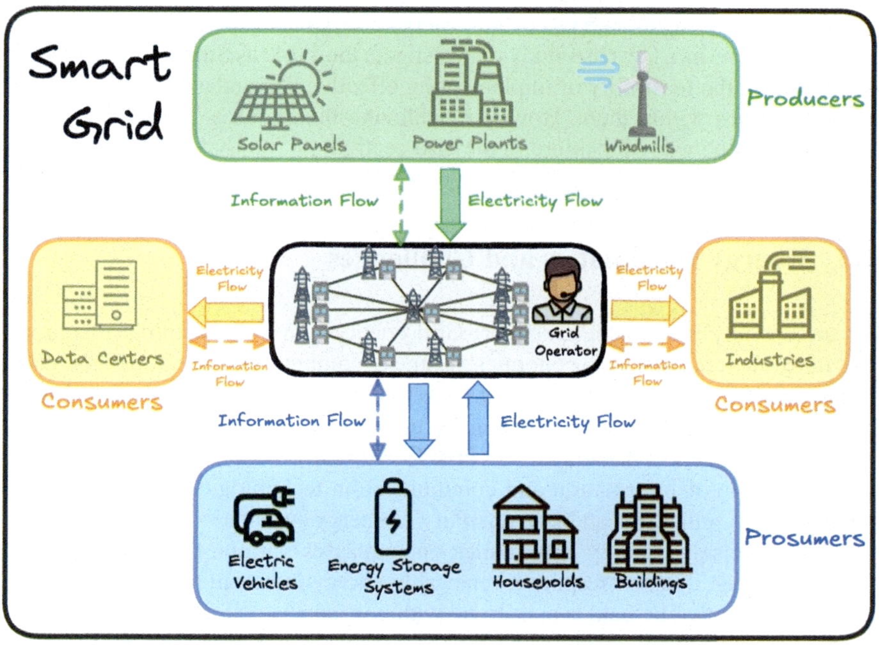

Fig. 1.2 A simplified graphical demonstration of Smart Grid, which is a typical example of a human-centered cyber-physical system. The grid operator (i.e., a human being) is the one who is responsible for managing the grid infrastructure. For control purposes, all the sub-components (e.g., electric vehicles, industries, etc.) exchange information with the central managing system using ICT (dotted lines). The green, orange, and blue regions illustrate the generating, consuming, and prosuming sub-systems respectively. The arrows demonstrate the flow of electricity to and from each of the three sub-components

Figure 1.2 provides a simplified overview of the Smart Grid as an HCPS (i.e., GO is the human involvement), illustrating some of its constituent sub-systems interconnected through ICT for information exchange as well as control. The sub-systems depicted in the green region represent examples of generation components, encompassing both conventional sources (e.g., nuclear power plants) and renewable sources (e.g., solar panels, windmills). The sub-systems in the orange region represent consumption entities like households, data centers, and industries. Meanwhile, the blue region represents sub-systems that can simultaneously consume and generate electricity. For instance, electric vehicles (EVs) consume energy while charging (i.e., grid to vehicle or G2V), but can also feed electricity back to the grid in certain cases (i.e., vehicle to grid or V2G) using their battery. Similar examples include households and buildings equipped with rooftop solar panels or energy storage systems (e.g., batteries).

On one hand, as highlighted in Sect. 1.1, our daily digital activities—primarily centered around computation and communication—are fundamentally supported by data centers. On the other hand, these facilities are substantial power demanders, drawing significant energy from society's overall energy grid. Consequently, this book will utilize data centers as a key case study to investigate their role as Smart Grid consumers and to assess the feasibility of implementing effective demand and energy management strategies within them. However, achieving this requires a multidisciplinary approach, which we will explore in the next section.

1.3 Energy Informatics and Challenges

To successfully implement the Smart Grid concept, an interdisciplinary approach that integrates engineering, economics, business, computer science, and other related disciplines is crucial. This book will explore foundational principles from a computer science viewpoint—while considering the use case of data centers in Smart Grid—introducing readers to the dynamic field of Energy Informatics. This area focuses on the application of information and communication technologies (ICT) to improve the efficiency, reliability, and sustainability of energy systems.

More precisely, Energy Informatics encompasses various domains, including smart metering, demand response, renewable energy integration, and energy management systems. The field is concerned with how data is collected, processed, analyzed, and utilized to optimize energy production, distribution, and consumption. For this purpose, this area leverages well-established computer science principles and methodologies (e.g., Markov chains and queuing theory) to address challenges within Smart Grid and energy systems.

As previously mentioned, central to Energy Informatics is the collection and management of vast amounts of data from various sources, such as smart meters, sensors, and IoT devices. This data can include real-time energy usage, generation patterns, and environmental conditions. Consequently, advanced analytics, including machine learning and statistical methods (e.g., ARIMA), are employed to model energy consumption patterns, forecast demand, and identify trends. For instance, predictive analytics can enable utilities to anticipate peak demand periods, allowing for better resource allocation. With the help of the derived models, Energy Informatics involves developing algorithms that optimize energy use across various scales. For example, algorithms can be designed to minimize energy costs while ensuring grid stability and maintaining service quality. These algorithms help stakeholders, such as utilities, businesses, and consumers, make informed decisions regarding energy usage and investments in energy technologies. However, several significant challenges must be addressed:

1. System architecture design: Creating a comprehensive system architecture that accommodates diverse use cases—such as electric mobility, data centers, smart homes, and smart cities—requires interoperable solutions. This is complex due to varying objectives among subsystems, different communication technologies (like wireless versus wired), and the diversity of existing infrastructure. SGAM was proposed to achieve such an interoperable architecture design for Smart Grid.
2. Optimization algorithms development: It is essential to develop algorithms for optimization, regulation, and control that consider multiple competing objectives. For instance, these algorithms must balance the goals of reducing energy consumption while enhancing performance across various time scales, such as seconds, minutes, and quarters of an hour.
3. Information processing: Achieving efficient, reliable, and robust information processing involves applying analytical methods grounded in physical laws—such as understanding that power equals current times voltage—as well as more modern, data-driven AI techniques. One example is predictive modeling, where forecasts are generated based on historical data related to a dependent variable and its associated features.
4. Safety and security analysis: It is vital to investigate potential safety and security threats through thorough modeling, analysis, and risk assessment, ensuring that the Smart Grid operates securely and reliably.

Tackling these challenges is essential for establishing a more efficient and resilient Smart Grid infrastructure. Throughout this book, we will illustrate how several of these critical challenges are addressed specifically within the context of data centers and the Smart Grid. We will explore the application of Energy Informatics to effectively manage power demand and optimize energy consumption in data centers. Ultimately, the insights and methodologies presented herein will empower students, researchers, academicians, and practitioners (e.g., stakeholders) to drive innovation and foster a more sustainable energy landscape.

1.4 Organization of the Book

This book is designed to offer a thorough and nuanced exploration of demand and energy management, catering to a wide audience with varied backgrounds and interests. Whether readers come from academia—such as professors and students—or from the industry, including engineers and practitioners, the content is structured to resonate with their specific needs and perspectives.

To effectively illustrate these concepts, the book centers on data centers as a pivotal use case within the Smart Grid framework. Data centers are not only significant consumers of energy but also pioneers in adopting innovative energy management practices. By examining real-world applications and case studies related to data centers, we aim to bridge the gap between theoretical knowledge and practical implementation. Ultimately, our goal is to equip readers with the insights and methodological foundations necessary to navigate the complexities of energy management in today's rapidly evolving technological environment, promoting a more efficient and sustainable future.

The first part of this book introduces the Smart Grid concepts and foundational principles of demand-side management (DSM). Chapter 2 begins by explaining the electricity chain stages: generation, transmission, distribution, and consumption. It categorizes power plants as base, medium, and peak-load, discussing their pros and cons. The transition from conventional grids to Smart Grid is explored, focusing on the SET Plan objectives and the Smart Grid architecture model (SGAM), which illustrates interoperability across layers, domains, and zones. The chapter emphasizes the role of information and communication technologies (ICT) in the energy transition and promotes the shift from "supply follows demand" to DSM, highlighting the need for a comprehensive management system for future energy networks. Chapter 3 provides a detailed overview of DSM in the Smart Grid context, outlining its benefits for consumers and utilities. It introduces concepts like shifting, shedding, and valley-filling for managing energy demand and discusses the timescales, pricing, and incentive schemes involved in DSM operations. Key components for implementing DSM—prediction models, Smart Meters, communication infrastructure, and hardware support—are emphasized, alongside a comparison of analytical and data-driven prediction techniques. The chapter also addresses communication requirements for Smart Meters and discusses remotely controllable hardware, including real-life application examples.

The second part of this book examines flexibility and DSM in data centers, highlighting their implementation through an energy management system (EMS). Chapter 4 focuses on the significant role of ICT in data centers, detailing essential resources such as servers, storage, networking equipment, cooling systems, and backup power systems. It addresses the feasibility of energy and demand management, introducing power usage effectiveness (PUE) as a metric for assessing efficiency. The chapter also discusses demand flexibility, differentiating between time and amount flexibility, and how components like workloads and cooling systems can adjust energy use. Chapter 5 explores the feedback control loop process in human-centered cyber-physical systems (HCPS), emphasizing the collaboration among hardware, operators, and stakeholders. It discusses human-machine interfaces for infrastructure reconfiguration and software orchestration of EMS, highlighting an agent-based approach and the contract net protocol (CNP). The chapter concludes by stressing the necessity of flexible agreements for effective demand adaptation and successful DSM practices in data centers, ensuring energy consumption is optimized while maintaining IT customer service level agreements (SLAs).

The third part of this book presents methodologies for developing power prediction models for servers in data centers and analyzing the power-performance trade-off, where enhancing server performance competes with reducing power demand and energy consumption. Chapter 6 highlights the importance of CMOS technology in microprocessors, tracing its evolution and advancements that boost power efficiency. It introduces analytical models for estimating power demand in CMOS circuits during idle and active states and discusses the von Neumann architecture, focusing on essential components like the CPU and RAM. The chapter emphasizes that servers are significant power demanders as well as energy consumers in data centers and outlines methodologies for predictive models of their power demands, considering variables such as voltage and frequency. This foundational knowledge aids in optimizing energy use and performance, particularly with dynamic voltage and frequency scaling (DVFS) technology. Chapter 7 explores how hardware manufacturers utilize DVFS to balance power demand and performance. It details DVFS governance rules for adjusting processor frequencies and presents mathematical metrics for evaluating performance needs. The chapter then discusses queuing theory and Markovian analysis for understanding the power-performance trade-off, establishing the single-server M/M/1 queuing model's relevance for this assessment. Finally, it introduces discrete-time Markov chain models for utilization-based governors, focusing on onDemand and Conservative strategies, and discusses steady-state probabilities to enhance understanding of these governance approaches.

In the final chapter, the author shares personal insights on the role of data centers in today's world, referencing a Bloomberg article that highlights challenges faced by grid and data center operators. The chapter explores emerging technologies such as renewable energy, edge computing, ARM technology, and AI-driven energy management systems (EMS) as solutions to enhance efficiency and sustainability. The author concludes by pondering whether data centers are a double-edged sword, essential for our digital lives yet raising concerns about their environmental impact and the need for balance and responsibility.

Part I
Foundations of Smart Grid and Demand-Side Management

The first part of this book covers the concepts of Smart Grid and the fundamental principles of demand-side management. Chapter 2 kicks off by elucidating the crucial stages of the electricity chain: power generation, transmission, distribution, and consumer usage. It then delves into the classification of power plants as base, medium, and peak-load plants, outlining their advantages and disadvantages. The evolution from the conventional grid to the Smart Grid is examined, with a focus on the objectives of the SET Plan. The Smart Grid architecture model (SGAM) is elaborated upon, highlighting its layers of interoperability, domains, and zones through illustrative examples. The significance of information and communication technologies (ICT) in the energy transition and Smart Grid operations is emphasized. The shift from the traditional approach of "supply follows demand" to the advocacy for "demand-side management" is promoted, along with the necessity of a comprehensive management system to supervise diverse energy networks in future energy systems.

Chapter 3 provides a comprehensive overview of demand-side management (DSM) within the context of Smart Grid, highlighting the benefits for both consumers and utility companies. It introduces practical concepts like shifting, shedding, and valley-filling to manage energy demand effectively. The chapter discusses various timescales used in DSM operations and explains pricing and incentive-based schemes. It emphasizes the four essential components for implementing DSM in Smart Grid: prediction models, Smart Meters, communication infrastructure, and hardware support. The chapter also compares analytical and data-driven techniques for prediction models and explores communication requirements for Smart Meters. Additionally, it delves into bidirectional communication interfaces and database technologies for meter data management. Lastly, it discusses remotely controllable hardware and provides real-life examples of their applications.

Chapter 2
Power and Smart Grid

2.1 Electricity Chain

2.1.1 Supply Follows Demand Paradigm

Electricity, much like food, water, or fuel, is an indispensable everyday necessity for our modern society. In developed nations, it is expected that electricity will be accessible at all times in areas where people reside. Regrettably, we often fail to appreciate the fact that the production and distribution of electricity are both costly and intricate endeavors. Consequently, we tend to consume electricity without fully acknowledging the expense and complexity involved in its entire chain. The anthropological perspective on the influence of electricity on modern society can be explored in Winther (2008).

Electricity is first generated at power plants, which are usually located far away from where it is consumed. This means that electricity must travel long distances before reaching society for consumption. In order to minimize transmission losses, the voltage of the generated electricity is increased through step-up transformers, typically from 110 to 765 kV. This is done in the transmission grid, which is also referred to as the high-voltage grid. The next step in the electricity chain is the distribution of electricity to society through the distribution grid.

To accomplish this, the voltage is reduced using step-down transformers to values usually around 50 kV or lower. The distribution grid is also known as the medium-voltage grid. Finally, once the electricity reaches society through distribution or feeder lines, transformers at the poles further reduce the voltage to 1 kV and 220/110 V before it enters factories and houses respectively. This grid is commonly referred to as the low-voltage grid. Figure 2.1 gives a brief graphical demonstration of the four distinct stages involved in the electricity chain. To gain a comprehensive and in-depth understanding of the electricity chain and its underlying components, readers are encouraged to consult (Bush 2014).

R. Basmadjian, *Demand-Side Energy and Power Management in Data Centers*, Lecture Notes in Energy 1, https://doi.org/10.1007/978-3-031-85956-4_2

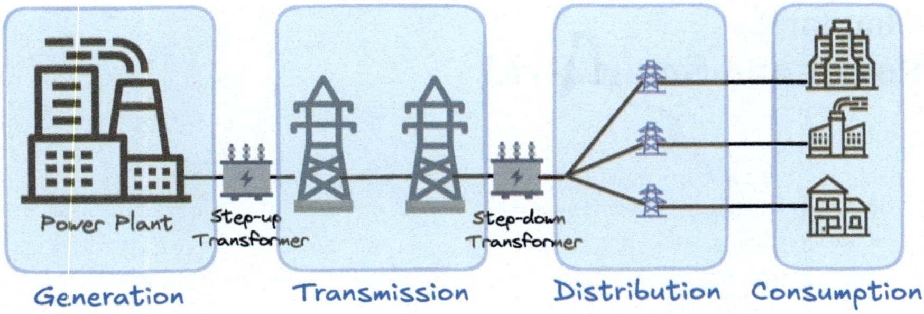

Fig. 2.1 A simplified graphical representation illustrating the four distinct stages of the electricity chain, starting from generation and ending with consumption

This electricity chain is key to how the legacy grid operates and the way electricity is provided to society. It is characterized by a stable power grid with consistent voltage levels (at high, medium, and low-voltage grids) and minimal fluctuations in frequency. Notably, stable European grids like those in the United Kingdom, Germany, and France experience very few blackouts. Maintaining a balance between electricity generation and demand (e.g., by the consumers) is essential to prevent blackouts. Traditionally, the belief has been that supply should align with demand. In other words, as demand increases, generation should also increase, following the "supply follows demand" paradigm. To understand how the legacy grid achieved this objective, it is crucial to provide an explanation of the different types of power plants and each type's characteristics in terms of response time and cost.

2.1.2 Types of Power Plants

Traditionally, power plants can be categorized into three types: base-load, medium-load, and peak-load. The primary purpose of base-load power plants is to meet the fundamental electricity needs. These power plants generate a constant amount of electricity but lack the ability to quickly adjust their generation in response to changes in demand. However, base-load power plants are the most cost-effective option and include examples such as nuclear, coal, and biomass plants.

Since base-load power plants alone cannot fully meet society's electricity demands, medium-load power plants are utilized to generate the additional electricity required. These power plants have a moderate level of responsiveness to changes in demand and offer affordable electricity prices. Examples of medium-load power plants include hydroelectric and renewable energy sources like photovoltaic (short PV) and windmills.

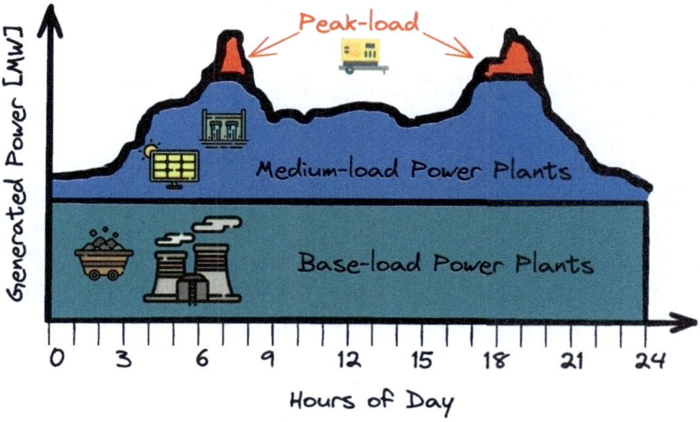

Fig. 2.2 Base-load power plants ensure a consistent supply of electricity, while medium-load ones adjust their generation based on demand. Peak-load power plants, on the other hand, rapidly generate electricity in response to sudden spikes in demand, such as during the hours between 6:00 and 9:00 AM as well as 6:00 and 9:00 PM depicted in the figure

However, there are certain periods, such as very hot days when air conditioning usage surges, that result in sudden peaks in electricity demand. During these periods, peak-load power plants come into play to generate electricity and match the sudden increase in demand. Peak-load power plants are characterized by their ability to react quickly, often within seconds. Yet, they are considerably expensive and, in some cases, not environmentally friendly, such as diesel generators. For a more comprehensive understanding of the various types of power plants, along with their specific characteristics and associated costs, interested readers can refer to Kaplan (2011). Figure 2.2 illustrates graphically the three main types of power plants and their involvement in the electricity generation.

The legacy grid effectively maintains a balance between supply and demand through autonomous mechanisms, relying on frequency control and without the need for communication. Additionally, as previously mentioned, highly developed European power grids experience minimal to no blackouts. These grids operate by generating electricity from large centrally managed power plants, which then flows through transmission and distribution grids until it reaches various sectors of society, including industries and residential areas as depicted in Fig. 2.1. Despite the apparent success of this system, it is important to explore the challenges present within the legacy grid and understand the factors that have driven its transformation into the Smart Grid.

2.2 Evolution from Legacy to Smart Grid

2.2.1 *Strategic Energy Technology Plan*

The transition from the legacy grid to the Smart Grid primarily occurred in Europe as a result of the European Commission's Strategic Energy Technology Plan (SET Plan) initiative, which was introduced in March 2007. The primary objective of this plan was to expedite the development of clean and sustainable energy sources, particularly renewables, while minimizing our reliance on fossil fuel-based technologies for electricity generation. The outlined plan sets forth three primary objectives to be accomplished by 2020 (Skjærseth and Eikeland 2023):

- A 20% reduction in the European Union's greenhouse gas emissions compared to the levels recorded in 1990.
- A 20% increase in the utilization of renewable energy sources for electricity consumption across all EU countries.
- A 20% improvement in energy efficiency compared to the values observed in 1990.

Following this SET Plan, numerous European countries have begun implementing it, by integrating renewable energy sources (RES) into their grid infrastructure. As an example, in the fall of 2010, Germany, inspired by the SET Plan, established an even more ambitious target. They aimed to achieve a 30% share of renewable energy sources by 2020 and a 50% share by 2030. This initiative gave rise to the Energy Transition, also known as Energiewende in German (Kühne and Weber 2018; Radtke and Canzler 2018), which was introduced in the spring of 2011. The primary objective of Energiewende was to replace nuclear power plants with renewable energy sources by 2022. It is worth noting that in 2010, nuclear power plants accounted for approximately 25% of the total electricity generation in Germany.

Figure 2.3 displays boxplots representing the percentage of RES (Basmadjian and Shaafieyoun 2023). Notably, Germany has already surpassed the 30% target set for 2020. The average percentage for the years 2019–2024 falls within the range of 40–60% throughout the entire year, where the maximum values are about 90% achieved during specific periods of the year 2023. From 2021 onwards, there has been a consistent and noticeable rise in the proportion of RES in Germany. This upward trend continues until reaching its peak in 2024. Figure 2.4 illustrates that the average share of renewable energy from 2019 to 2024 stands at an impressive 50%. This statistic confirms that Germany has already met the ambitious target outlined in its Energy Transition plan for 2030!

Thus, the integration of renewable energy sources was the driving factor of the transition from the legacy grid to the Smart Grid, especially in Europe. In this respect, information and communication technologies (ICT) played a vital role (see Sect. 2.3) in improving the efficiency and intelligence of various aspects of power generation, transmission, distribution, consumption, as well as accounting and billing processes.

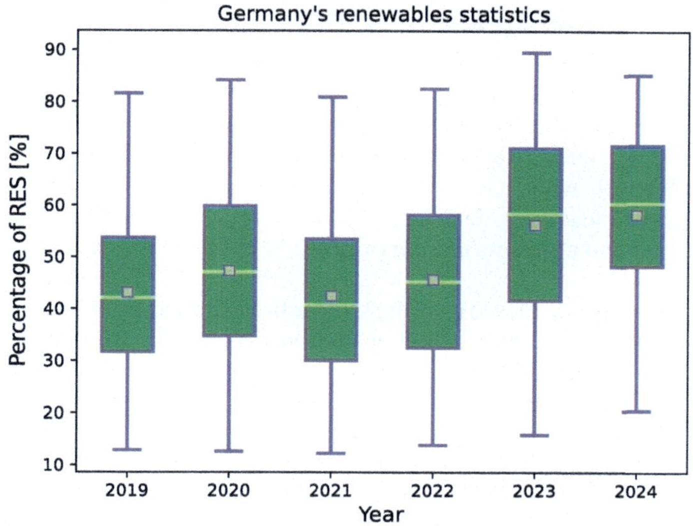

Fig. 2.3 Percentage of Germany's renewable energy sources (RES) from 2019 to 2024 (Basmadjian and Shaafieyoun 2023). The horizontal green lines and square boxes denote the median and mean respectively

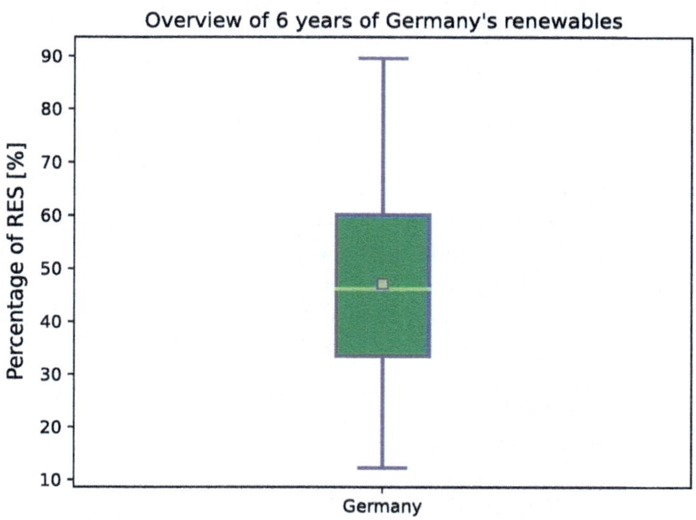

Fig. 2.4 Percentage of Germany's renewable energy sources (RES) of all six years concatenated. The horizontal green lines and square boxes denote the median and mean respectively, showing an average of 50% of renewables when considering all the six years

2.2.2 *Interoperability and SGAM*

When designing and developing new systems and concepts like the Smart Grid, ensuring interoperability is a crucial factor. Interoperability refers to the ability of different systems, devices, or applications to communicate, exchange data, and work together effectively without encountering compatibility issues. It ensures that diverse technologies can seamlessly interact and function in a coordinated manner, enabling users to access and utilize various services and resources across different platforms or networks.

Interoperability is achieved by using standardized interfaces. For instance, universal serial bus (USB) and high-definition multimedia interface (HDMI) are common examples of standardized interfaces used in our daily activities with personal computers (PCs). When devices have these standardized interfaces, they can exchange information regardless of the differences in hardware (like PCs, TVs, mobile phones) or software (such as Android, Linux, and Windows). For example, a Windows-based PC can be connected to an Android-based TV thanks to the HDMI interface. This promotes connectivity, collaboration, and integration among different systems, leading to improved efficiency, productivity, and user experience.

As highlighted in Sect. 1.2, the Smart Grid is described as a human-centered cyber-physical system (HCPS). As such, it is a complex system (see Fig. 1.2) composed of various sub-components, each with its unique characteristics in terms of hardware (such as heterogeneous infrastructure), software (including diverse operating systems and implementations), and communication technologies (such as wireless options like LTE and ZigBee, as well as wired options like PLC and Ethernet). Therefore, the notion of interoperability is particularly emphasized in the context of Smart Grid, due to the aforementioned heterogeneity involved in hardware, software, and communication technologies.

In response to the interoperability challenges within the context of Smart Grid, the European Commission introduced the Smart Grid architecture model (SGAM) (Smart grid reference architecture 2012) as a proposed framework to aid in the design and development of such a concept. This model consists of five interoperability layers, each comprising two dimensions of zones and domains. The five layers, from top to bottom, are Business, Function, Information, Communication, and Component.

The five domains represent different stages in the electricity chain introduced in Sect. 2.1, including Generation, Transmission, Distribution, Distributed Energy Resources (DER), and Customer Premises. Additionally, the six zones are Process, Field, Station, Operation, Enterprise, and Market. The SGAM interoperability layers can be described in the following order:

1. The Business layer: It focuses on clearly defining the business goals, objectives, use cases, and scenarios of the Smart Grid system. It is important to provide detailed descriptions to avoid any misunderstandings. An example of a use case can be "use solar energy during the daytime".

2. The Function layer: In this layer, the specific functions required to achieve the goals, objectives, use cases, and scenarios defined in the Business layer are specified. An example of a function for the aforementioned use case can be "power generation prediction of PV" based on weather forecasts.
3. The Information layer: It is responsible for defining the necessary data models and structures that the functions in the Function layer require for communication and information exchange. For instance, this can be a data model for the above power prediction function consisting of temperature, humidity, location of the solar panel, global horizontal irradiance (GHI), etc.
4. The Communication layer: Here, the communication protocols are specified for the exchange of information identified in the Information layer. Typically, these protocols are based on TCP over secure communication channels like SSL.
5. The Component layer: It describes the hardware infrastructure of the Smart Grid, including power plants, transformers, feeder lines, solar panels, and customer equipment such as washing machines and refrigerators, to name it few.

The five layers, illustrated in Fig. 2.5, collaborate to ensure the interoperability and efficient functioning of the Smart Grid system. Within the SGAM framework, the six zones have distinct roles. The Process zone represents the smallest entities in the system responsible for generating or consuming electricity. For example, in the DER and Customer Premises domains, entities like a windmill and a washing machine fall under this zone, respectively.

Multiple Process entities are consolidated in the Field zone, which may encompass several windmills in the DER domain or all the appliances within an apartment in the Customer Premises domain. The Station zone is created by combining multiple

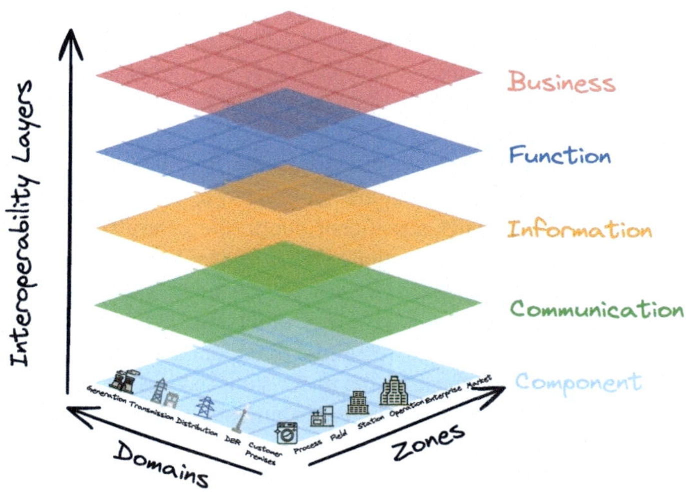

Fig. 2.5 The SGAM's five interoperability layers, with each layer's corresponding zones and domains (Smart grid reference architecture 2012)

Field entities. In the Customer Premises domain, this could be a building comprising several apartments (Field zone), each equipped with different appliances (Process zone).

The Operation, Enterprise, and Market zones, which are the remaining three zones, represent higher-level logical groupings typically implemented using software solutions. For instance, in the Operation zone within the Customer Premises domain, a building management system (BMS) is utilized as a software-based solution for managing multiple buildings.

Another example within the Operation zone is the charging station management system (CSMS) used for overseeing the charging stations for electric vehicles. In Kirpes et al. (2019), an illustration is provided on the application of the Smart Grid architecture model (SGAM) in the context of electric mobility, known as the electric mobility system architecture (EMSA) model. The utilization of SGAM in other cyber-physical systems, such as Industry 4.0, along with the fundamentals of the standard IEC 62559 use case methodology and its relationship to SGAM, is discussed in Gottschalk et al. (2017).

2.3 Importance of ICT in Smart Grid

According to Momoh (2012), the Smart Grid concept can be described as a network that has self-healing capabilities and utilizes dynamic optimization techniques. These techniques make use of real-time measurements to minimize network losses, ensure voltage stability, enhance reliability, and improve asset management. From this definition, it is evident that information and communication technologies (ICT) play a vital role in enabling the implementation of the Smart Grid concept. This is further supported by the presence of dedicated Information and Communication layers in the Smart Grid architecture model (SGAM) as shown in Fig. 2.4. However, it is important to acknowledge that ICT, including the Internet, has known vulnerabilities in terms of cybersecurity.

When ICT is introduced and integrated into the Smart Grid, the cybersecurity concerns associated with the Internet are transferred to this safety-critical system. This vulnerability is particularly significant due to the centralized structure of the Smart Grid and the inadequate utilization of security technologies. For readers seeking insights into security and privacy issues and challenges within the Smart Grid context, references (Khurana et al. 2010; McDaniel and McLaughlin 2009) provide valuable information. Therefore, it is essential to understand the necessity of ICT in the Smart Grid and carefully consider the underlying risks associated with it.

Indeed, renewables are undeniably environmentally-friendly energy sources, making them an ideal alternative to fossil fuels. However, they do face three significant challenges (Bird et al. 2013): mismatch, uncertainty, and decentralization. Mismatch occurs when the demand for energy does not align in timescale with the power generation curve. For example, we may choose to do laundry during a time when power generation from solar panels is expected to occur later. Additionally, certain renewable sources, like solar panels, experience a period of minimal electricity generation during late evening until early morning, commonly referred to as "dead calm". These issues pose additional complications in effectively harnessing renewable energy.

Secondly, despite advancements in prediction models, particularly those utilizing machine learning and artificial intelligence, there remains a degree of uncertainty in accurately forecasting the generation of renewable energy sources. Even with state-of-the-art prediction models boasting an accuracy rate of 90–95%, the remaining 5% of inaccuracy in predicting renewable energy generation necessitates significant investments in backup generators from conventional power plants. Consequently, this situation could potentially result in an escalation of electricity prices and a reliance on environmentally harmful alternatives such as diesel generators, ultimately defeating the purpose of transitioning to renewable energy sources. A comprehensive review of the impact of forecasting errors on electricity prices from the perspective of solar panels can be found in Gandhi et al. (2024).

Lastly, decentralization poses challenges at the low-voltage grid level, primarily caused by the installation of renewable energy sources like solar panels on residential rooftops. Apart from self-consumption, these sources can also feed electricity back into the low-voltage grid. However, unregulated feed-in can result in voltage spikes, leading to fluctuations and potential blackouts. Additionally, the integration of electric vehicles into urban areas further exacerbates the issue. Uncontrolled charging of these vehicles at the low-voltage grid can cause sudden voltage drops, potentially resulting in blackouts. For further information on this and other challenges related to electric mobility, readers are advised to refer to Eider et al. (2017).

In the coming years, there will be an abundance of energy available. However, the difficulty lies in aligning our society's electricity demand with the actual generation of energy, in the presence of the aforementioned three challenges related to renewables. To overcome this hurdle, a smart and intelligent energy management system (EMS) will be necessary to address autonomously as well as with minimum human intervention those challenges to some extent or in their entirety.

The corresponding EMS is a software-based solution that requires predictions (e.g., prices, generation, demand) to make the most suitable decision. Furthermore, the devised algorithms of such an EMS run on computing systems such as servers or IoT devices, and require information coming from the installed sensors. Hence, it becomes evident that without the integration of information and communication technologies (ICT), achieving effective energy management would be arduous, if not impossible!

2.4 Paradigm Change to Demand-Side Management

In Sect. 2.1.1, it is stated that maintaining a steady balance between demand and supply is crucial for the power grid. Any mismatch between the two can result in voltage and frequency fluctuations, ultimately leading to a blackout. The traditional power grid operated under the assumptions that (1) society's demand for electricity cannot be controlled, (2) electricity cannot be stored within the grid, (3) electricity generation is centralized through large power plants located near the high-voltage transmission grid, and (4) society's electricity demand is solely met by the low-voltage grid.

The legacy grid was originally designed to maintain a balance between supply and demand through frequency control, adhering to the principle of "supply follows demand." This means that, as electricity demand from society rises, the system is intended to adjust the supply of electricity in response to that demand. As depicted in Fig. 2.6, this paradigm showcases a unidirectional flow of electricity: it travels from generation sources directly to end consumers. In this traditional setup, power plants respond to the varying needs of consumers, ensuring that supply meets demand at all times.

Section 2.2 discusses how the legacy grid incorporated renewable energy sources. However, as highlighted in Sect. 2.3, renewables present challenges such as mismatches in timescale, uncertainty in predictions, and decentralization. It became evident that the legacy grid alone could not accommodate these renewables without the assistance of ICT, which play a crucial role in addressing the aforementioned challenges. As a result, the shortcomings of the "supply follows demand" paradigm became apparent, leading to its replacement with a new paradigm called "demand-side management" (also known as DSM).

Figure 2.7 presents a comprehensive overview of the Smart Grid concept, illustrating the bidirectional flow of both power and information in the DSM paradigm. This means that power and information can move from the generation source to customer premises, as well as in the opposite direction. It is worth noting that the flow of information relies on the use of communication infrastructure (e.g., ICT). As a result, consumers can have solar panels installed on their rooftops to supply electricity back to the grid, and consequently become prosumers (i.e., consumers who are also producers).

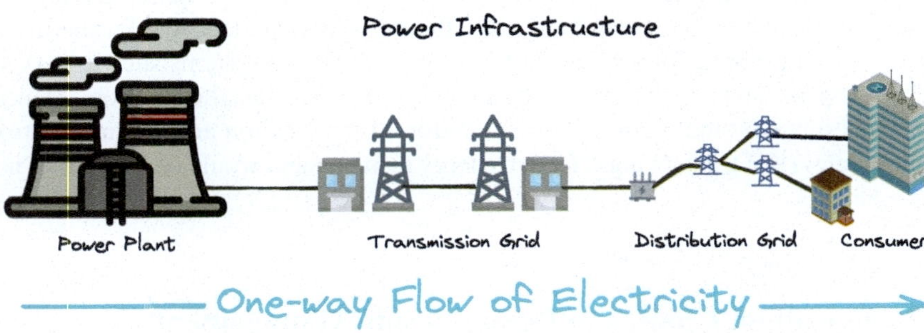

Fig. 2.6 In the "supply follows demand" paradigm, the flow of power is unidirectional, starting from the generation source and moving towards the consumers

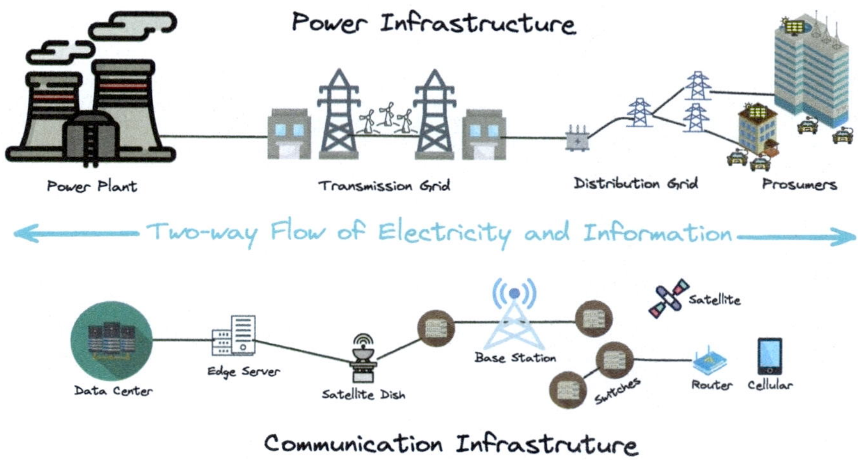

Fig. 2.7 In the paradigm of "demand-side management", both the flow of power and information are bidirectional. This means that in this approach, electricity can be generated at the customer premises (e.g., solar panels on the rooftop), allowing for a two-way flow of electricity between the consumers and the grid

Additionally, due to the bidirectional information flow, grid operators can inquire about the demand and consumption of their consumers (provided they have a Smart Meter), and consumers can request status-related information from their grid operators, such as electricity prices or forecasts of the next day's generation from renewables. Further details on this topic will be covered in Sect. 3.4. For a thorough understanding of the technologies and services related to the communication infrastructure of Smart Grid, it is recommended that readers refer to Sendin et al. (2016).

To summarize, the core concept of DSM involves the identification and utilization of the inherent flexibilities of consumers (Gellings and Chamberlin 1987). To accomplish this, two highly effective approaches, namely load shifting and shedding (see Sect. 3.2), enable decentralized management on the demand side. It is advantageous for intelligent and smart energy management systems to autonomously carry out this management, thereby enhancing the reliability of power supply, even in the presence of intermittent and fluctuating renewable energy sources. Further information on the benefits and challenges of DSM can be found in Strbac (2008).

2.5 Future Energy Management

Over time, various independent networks have been developed and incorporated into our communities. Key networks include electricity, transportation, district heating, and gas networks as shown in Fig. 2.8. Each of these networks is managed separately

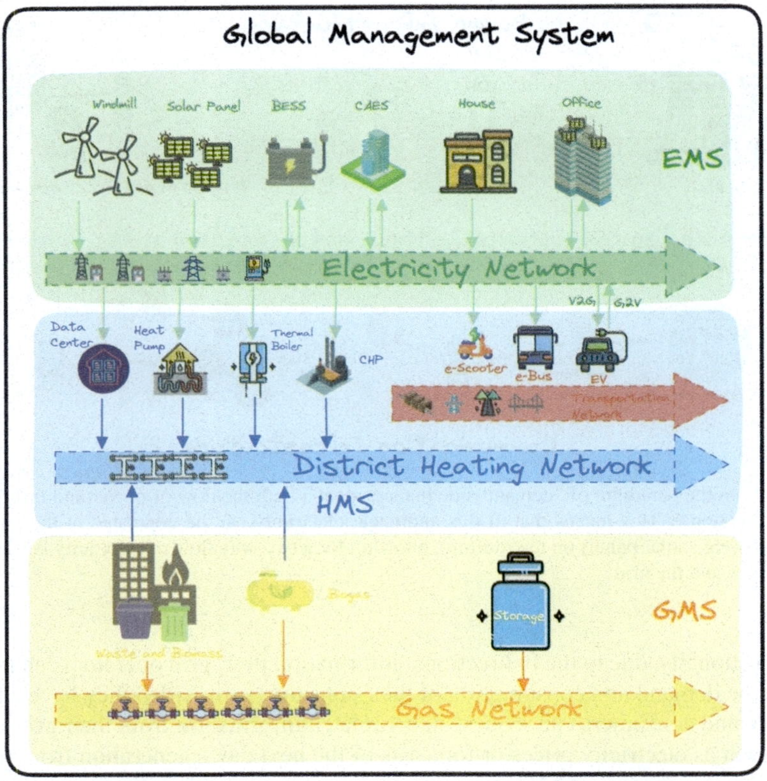

Fig. 2.8 A simplified representation of a global management system that effectively orchestrates and controls all the underlying networks to ensure a balance between supply and demand. To achieve this system, it is necessary to have interoperable and standardized interfaces that allow for seamless communication and integration. The arrows illustrate the flow of assets: green, blue and orange for electricity, heating and gas respectively

and has its own established infrastructure. For example, within the electricity network (i.e., the power grid), sources like solar panels and windmills contribute electricity to the grid.

There is a growing trend of integrating energy storage systems like batteries into the grid. Those systems are particularly beneficial for renewable energy as they enable the storage of surplus energy for later use, such as feeding electricity back to the grid. Unlike solar panels and windmills that only generate electricity, these energy storage systems (ESSs) are known as prosumers as they can both consume (e.g., charge the battery) and produce (e.g., discharge the battery) electricity.

Two commonly used technologies for energy storage systems are batteries (referred to as BESS) and compressed air (referred to as CAES). These technologies function by storing energy in the form of battery cells for BESS and compressed air for CAES. During times of low energy demand, surplus electricity from renewables is utilized to charge batteries or compress air, storing it in batteries or specialized containers. When energy demand increases, the batteries are discharged, or the compressed air is released, either returning electricity directly to the grid for BESS or powering turbines for CAES to generate electricity.

Presently, batteries have a shorter lifespan compared to CAES, with batteries typically lasting 3–5 times less. Additionally, there is a concern regarding the risk of battery overheating leading to potential explosions, particularly in critical systems like the Smart Grid. Eventually, there is a possibility that the current reliance on batteries for energy storage may shift towards CAES if no new battery technologies, like solid-state batteries, are introduced to address the limitations and concerns of BESS (Alami et al. 2022). Note that buildings, residential houses, and industrial premises are connected to the electricity network as energy consumers, or prosumers in the case of solar panels installed at rooftops and/or BESS. The management of this network is realized using the energy management system (EMS).

Within the transportation network, various electric mobility options are available, including electric vehicles (EVs), e-buses, e-trains, e-bikes, and e-scooters, to name a few. These modes of transportation are connected to the electricity network so that they can be charged from the grid (referred to as G2V or grid-to-vehicle). Some of them have also the capability of feeding electricity back to the grid (referred to as V2G or vehicle-to-grid) (Das et al. 2020). The arrows in Fig. 2.8 illustrate the flow of electricity where a bidirectional arrow indicates the capability of charging and discharging as is the case for instance with some EVs or BESS. Readers interested in the benefits and challenges of the V2G technology can refer to Yilmaz and Krein (2012).

In terms of the district heating network, there are multiple methods of generating heat. Thermal boilers, heat pumps, and electric cartridges all utilize electricity to produce heat. Recently, there have been studies and PoC projects that demonstrate the possibility of transferring the excess heat generated by data centers and their IT infrastructure to the district heat network (Huang et al. 2020; Wahlroos et al. 2017). The technology known as combined heat and power (CHP) produces electricity for the grid while also utilizing the dissipated heat by injecting it into a heat network. The heat network is independently controlled and managed through a heat management system (HMS). Lastly, systems of varying scales, from small domestic stoves to large industrial power or heating plants, utilize the combustion of waste and solid biomass or biogas to generate both heat and gas. This network is managed by the gas management system (GMS).

Each network outlined earlier operates under its own management system, such as EMS, HMS, or GMS as depicted in Fig. 2.8. Looking ahead to future energy systems, there arises a necessity for these networks to seamlessly and autonomously interact and exchange assets like electricity, heat, and gas. This requirement underscores the need for an integrated, intelligent, and global management system, often in the form of a software solution. The primary goal of this management system is to autonomously balance the supply and demand of each network, even in the face of conflicting objectives, such as feeding back power from EVs to the electricity network without compromising the battery's health.

The aforementioned software-driven system should not only be adaptable to various scenarios, but also be interoperable to facilitate seamless interaction among the diverse technologies utilized in different networks, despite their inherent disparities. Consequently, there will be an increasing demand for standardized interfaces in the future, along with greater integration of the information infrastructure with distributed system intelligence, as well as for new technological advances such as direct conversion of power to gas and back to power.

These advancements will bolster the efficiency, flexibility, and stability of the interconnected networks. It is important to note that due to the inherent conflicting objectives of the different networks, finding the global optimal solution for the asset flow problem in this global energy management system is a complex task, as evidenced in La Scala et al. (2021).

2.6 Summary

This chapter commenced with an overview of the electricity chain, providing a comprehensive understanding of its four fundamental phases: power generation, transmission, distribution, and consumer consumption. It also delved into the classification of power plants into base, medium, and peak-load plants, highlighting the advantages and drawbacks of each type. The evolution from the traditional grid to the Smart Grid was then explored, emphasizing the goals of the SET Plan. The Smart Grid architecture model (SGAM) received significant attention, with a concise introduction to its five interoperability layers, domains, and zones, accompanied by relevant examples.

The chapter further underscored the indispensability of information and communication technologies (ICT) in achieving energy transition and illustrated the pivotal role it plays in the Smart Grid. The necessity for a paradigm shift from "supply follows demand" to "demand-side management" (DSM) was elucidated, citing the underlying reasons. Ultimately, the chapter emphasized the imperative need for an interoperable global management system that autonomously governs the diverse existing networks of electricity, transportation, district heating, and gas for future energy systems.

References

Smart grid reference architecture. Technical report, Smart Grid Coordination Group (2012)

Alami AH, Yasin A, Alrashid R, Alasad S, Aljaghoub H, Alabsi G, Alketbi L, Alkhzaimi A, Alteneji A, Shikhli S (2022) Experimental evaluation of compressed air energy storage as a potential replacement of electrochemical batteries. J Energy Storage 54:105263

Basmadjian R, Shaafieyoun A (2023) Assessing arima-based forecasts for the percentage of renewables in Germany: insights and lessons for the future. Energies 16(16)

Bird L, Milligan M, Lew D (2013) Integrating variable renewable energy: challenges and solutions. Technical report, National Renewable Energy Lab. (NREL), Golden, CO (United States)

Bush SF (2014) Smart grid: communication-enabled intelligence for the electric power grid. Wiley
Das S, Acharjee P, Bhattacharya A (2020) Charging scheduling of electric vehicle incorporating grid-to-vehicle (g2v) and vehicle-to-grid (v2g) technology in smart-grid. In: 2020 IEEE international conference on power electronics, smart grid and renewable energy (PESGRE2020), pp 1–6. IEEE
Eider M, Sellner D, Berl A, Basmadjian R, de Meer H, Klingert S, Schulze T, Kutzner F, Kacperski C, Stolba M (2017) Seamless electromobility. In: Proceedings of the eighth international conference on future energy systems, pp 316–321
Gandhi O, Zhang W, Kumar DS, Rodríguez-Gallegos CD, Yagli GM, Yang D, Reindl T, Srinivasan D (2024) The value of solar forecasts and the cost of their errors: a review. Renew Sustain Energy Rev 189:113915
Gellings CW, Chamberlin JH (1987) Demand-side management: concepts and methods
Gottschalk M, Uslar M, Delfs C (2017) The use case and smart grid architecture model approach: the IEC 62559–2 use case template and the SGAM applied in various domains. Springer
Huang P, Copertaro B, Zhang X, Shen J, Löfgren I, Rönnelid M, Fahlen J, Andersson D, Svanfeldt M (2020) A review of data centers as prosumers in district energy systems: renewable energy integration and waste heat reuse for district heating. Appl Energy 258:114109
Kaplan S (2011) Power plants: characteristics and costs. DIANE Publishing
Khurana H, Hadley M, Lu N, Frincke DA (2010) Smart-grid security issues. IEEE Secur & Priv 8(1):81–85
Kirpes B, Danner P, Basmadjian R, Meer HD, Becker C (2019) E-mobility systems architecture: a model-based framework for managing complexity and interoperability. Energy Inf 2:1–31
Kühne O, Weber F (2018) Bausteine der Energiewende. Springer
La Scala M, Bruno S, Nucci CA, Lamonaca S, Stecchi U (2021) From smart grids to smart cities: new challenges in optimizing energy grids, vol 2. Wiley
McDaniel P, McLaughlin S (2009) Security and privacy challenges in the smart grid. IEEE Secur & Priv 7(3):75–77
Momoh JA (2012) Smart grid: fundamentals of design and analysis, vol 63. Wiley
Radtke J, Canzler W (2018) Energiewende. Springer
Sendin A, Sanchez-Fornie MA, Berganza I, Simon J, Urrutia I (2016) Telecommunication networks for the smart grid. Artech House
Skjærseth JB, Eikeland PO (2023) Governing eu low-carbon innovation: From strategic energy technology plan to European green deal. In: Handbook on European union climate change policy and politics. Edward Elgar Publishing, Cheltenham, UK, pp 259-273
Strbac G (2008) Demand side management: benefits and challenges. Energy Policy 36(12):4419–4426
Wahlroos M, Pärssinen M, Manner J, Syri S (2017) Utilizing data center waste heat in district heating-impacts on energy efficiency and prospects for low-temperature district heating networks. Energy 140:1228–1238
Winther T (2008) The impact of electricity: development, desires and dilemmas. Berghahn Books
Yilmaz M, Krein PT (2012) Review of benefits and challenges of vehicle-to-grid technology. In: 2012 IEEE energy conversion congress and exposition (ECCE), pp 3082–3089. IEEE

Chapter 3
Demand-Side Management

3.1 Benefits and Potentials

Chapter 2 highlights the necessity of a paradigm shift from "supply follows demand" to "demand-side management" (DSM). The reason for this shift is that in the former approach, the infrastructure must be capable of meeting peak load demands whenever they occur. This often requires significant investments by grid operators in infrastructure that may only be utilized for a limited duration and incurs additional costs. For instance, a study cited as Basmadjian et al. (2016) illustrated that surpassing the established limit by the DSO of 61 MW even for 1 s incurred an additional cost of €54 per kW in 2011. This resulted in approximately €28,000 in extra expenses for the scenario analyzed, attributed to a total of 516 kW exceeding[1] the threshold that year.

Apart from the significant costs involved, the grid operator relies on environmentally unfriendly sources, as mentioned in Sect. 2.1, such as diesel generators, to meet the sudden surge in electricity demand. A more favorable approach would have been to manage the demand of society to decrease this peak. Another scenario where the "supply follows demand" paradigm reveals its limitations is when society's energy demand is low, and there is an excess of electricity from renewable sources. In this case, a more effective approach would be to encourage society to increase their demand, preventing the wastage of electricity generated from those green sources.

DSM offers a superior alternative to the traditional "supply follows demand" approach, particularly in the era of renewable energy sources. It represents a new paradigm with significant potential, as it (1) helps defer investments in new infrastructure by reducing peak demand, (2) promotes environmental conservation by minimizing the use of ecologically harmful fossil-based generators like diesel, (3) enhances system reliability by investing in diverse alternatives, and (4) empowers consumers to control and even reduce their energy bills. The potential of DSM is primarily based on several key factors such as:

---

[1] Mostly happened in the months of November (twice) and December (6 times).

R. Basmadjian, *Demand-Side Energy and Power Management in Data Centers*, Lecture Notes in Energy 1, https://doi.org/10.1007/978-3-031-85956-4_3

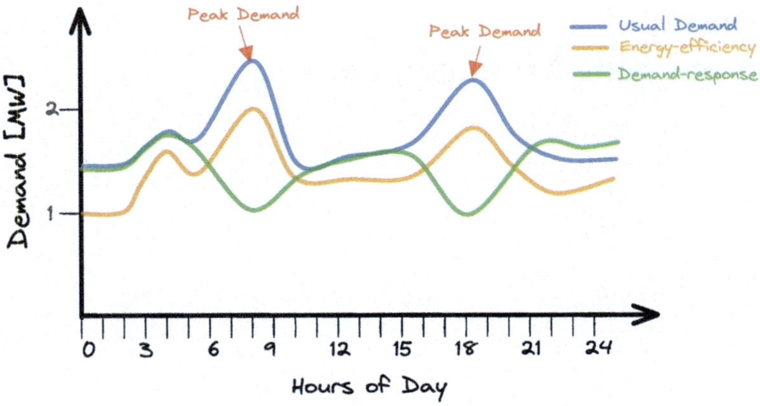

Fig. 3.1 Example of the society's demand for a workday (blue curve) with the two DSM programs of energy-efficiency (orange curve) and demand-response (green curve)

1. The availability and timing of information provided to consumers.
2. The start time, duration, and amount of power to adapt (e.g., reducing during shortages or increasing during excess of electricity).
3. The presence and performance of a digital metering infrastructure (e.g., Smart Meters).
4. The automation of consumer equipment (e.g., washing machines, air conditioners).
5. The electricity pricing schemes and contracts in place.

Utility companies were motivated to adopt DSM due to its numerous benefits and potential. It involves the identification of activities that can influence consumers' electricity usage and encompasses the planning, implementation, and monitoring of initiatives aimed at encouraging consumers to alter their electricity consumption patterns. Within the framework of DSM, two major programs are implemented: energy-efficiency (EE) and demand-response (DR). The primary objective of the EE program is to reduce overall energy consumption (measured in kilowatt-hours, kWh). On the other hand, the DR program focuses on managing the demand (measured in kilowatts, kW) in response to changes in supply conditions.

In Fig. 3.1, a visual representation is provided to illustrate the two programs in relation to the typical electricity demand of society. The example assumes that there are two peaks in electricity demand, occurring in the early morning between 7:00 AM and 9:00 AM, and in the evening between 6:00 PM and 8:00 PM. The orange curve represents the scenario where certain appliances, such as heat boilers and refrigerators, are energy-efficient and consume less energy. On the other hand, the green curve demonstrates the impact of demand-response, specifically peak demand reduction, achieved by shedding (e.g., removing) and shifting (e.g., rescheduling) electricity demand during peak times. It is important to note that the optimal solution would

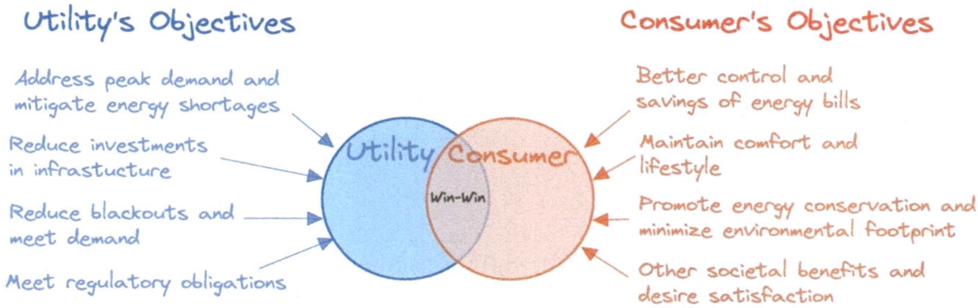

Fig. 3.2 The implementation of DSM results in a mutually advantageous scenario for both utilities and consumers. Consumer objectives are represented on the right in red, while utility goals are depicted on the left in blue. The benefits of DSM, labeled as "win-win," are found at the intersection of these two circles. Please note that consumers encompass various sectors such as industrial facilities, hospitals, residential areas, among others

involve a combination of both programs to achieve the best results for demand-side management (DSM).

Figure 3.2 showcases the advantages of DSM from the viewpoints of both consumers and utilities, illustrating the mutual benefits that can be achieved. For instance, consumers can achieve savings on their heating bill of 8–10% by simply reducing the temperature setpoint of their heating system by 1 °C (e.g., from 20 °C to 19), while still maintaining their desired comfort and lifestyle (Shann et al. 2017). Additionally, this allows consumers to have better control over their energy usage, leading to reduced energy bills. Moreover, by actively participating in DSM, consumers contribute to the shift towards ecological practices, as it helps minimize the environmental impact of greenhouse gas emissions.

It is important to note that consumer participation in DSM is voluntary, allowing them to align their actions with their own preferences and values. On the flip side, DSM enables utilities to avoid the need for extensive infrastructure expansion by effectively managing the increased demand of consumers (e.g., society). This approach allows utilities to address peak demand, minimizing the risk of blackouts and mitigating energy shortage situations. Additionally, DSM provides utilities with the means to meet regulatory requirements set by governing bodies, such as the implementation of initiatives like the SET Plan in Europe (refer to Sect. 2.2). It is crucial to recognize that households, factories, and various other societal entities play a significant role in electricity consumption, collectively referred to as consumers.

3.2 Concepts for Demand Management

The core principle of DSM revolves around harnessing the flexibility of loads, such as demand. To achieve this objective, three distinct concepts have been introduced to leverage the flexibility of load and effectively manage consumer demand. These concepts include load-shifting, valley-filling, and peak-shedding, which is also referred to as peak-shaving or clipping.

3.2.1 Load-Shifting

The primary goal of this concept is to shift consumer demand from on-peak to off-peak periods. By doing so, the total energy consumption for a specific time period, such as a day, remains unchanged. Shifting, in this context, refers to rescheduling activities to later times in order to reduce overall consumer demand. For example, if you have the flexibility to choose the day and time for doing laundry, it becomes irrelevant whether it is done on Monday or Tuesday, or at different times within the same day.

By opting to do laundry during off-peak hours, such as moving from 7 AM (peak time) to 12 AM (off-peak time), the objective is to alleviate demand during peak periods. In the provided example, Fig. 3.3 illustrates the process of shifting or moving two similar loads of 0.5 kW each (e.g., a washing machine) from on-peak hours (7:00 AM to 9:00 AM or 6:00 PM to 8:00 PM in Fig. 3.3a) to off-peak hours, specifically at 12:00 AM in Fig. 3.3b. It is important to note that the application of this technique keeps the overall energy consumption the same. Additionally, modern home appliances are equipped with an automated timer system. This system allows the equipment to start its operation without the need for human presence.

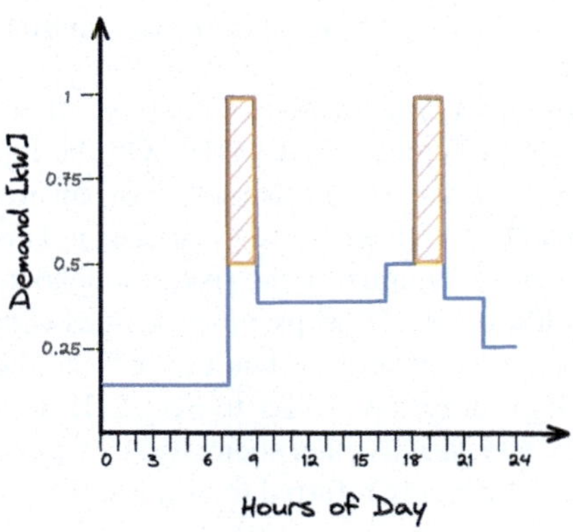

(a) Before load-shifting.

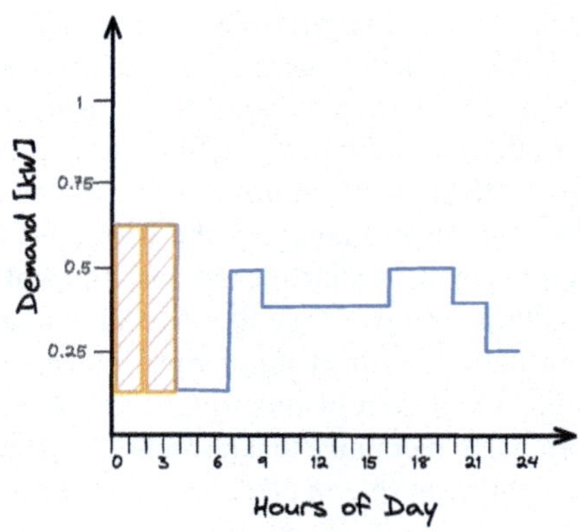

(b) After load-shifting.

Fig. 3.3 The graph depicts the load-shifting concept. The orange dashed rectangles represent two identical loads with a power demand of 0.5 kW, each lasting for 2 h. The illustration showcases the process of moving these loads from on-peak hours (between 7:00 AM and 9:00 AM and 6:00 PM to 8:00 PM) to off-peak hours (between 12 AM and 4 AM in this instance). In Fig. 3.3a, the total energy consumption is 9.75 kWh, and it remains unchanged in Fig. 3.3b after shifting the load

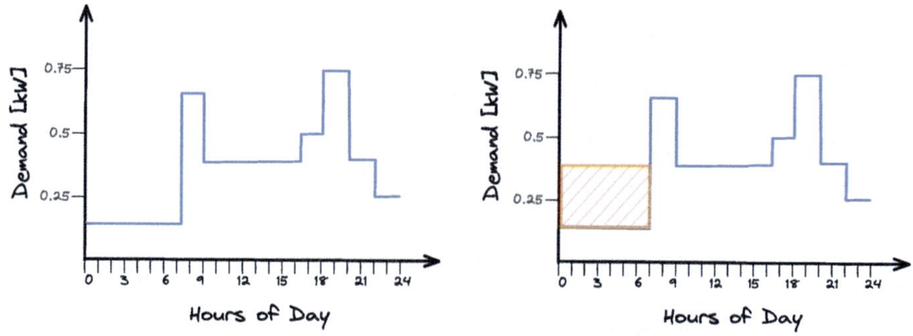

(**a**) Before valley-filling.

(**b**) After valley-shifting.

Fig. 3.4 The graph demonstrates the valley-filling concept. The orange dashed rectangle symbolizes the inclusion of extra load, such as charging a battery energy storage system (BESS), or storing energy thermally during off-peak hours (between 12:00 AM and 7:00 AM). In Fig. 3.4a, the total energy consumption for the entire day is 8.5 kWh, while in Fig. 3.4b, it increases to 10.25 kWh

3.2.2 Valley-Filling

The objective of this concept is to increase demand during off-peak hours, typically between 12:00 AM and 7:00 AM. This is done in order to proactively store energy, which can then be utilized to reduce peak demand throughout the day. Examples of this include charging electric vehicles (EVs), or battery energy storage systems (BESSs), as well as initiating the operation of water heaters or pumps to avoid their usage during peak times.

As a result, the valley-filling concept introduces new loads, potentially leading to an increase (1.75 kWh in the case of orange dashed rectangle leading to a total of 10.25 kWh in Fig. 3.4b) in the typical energy consumption (8.5 kWh for the given example in Fig. 3.4a) of that particular day. Please note that when looking at a shorter timeframe, such as a single day, the valley-filling concept may lead to an increase in energy consumption as it involves anticipating storage needs. However, when considering a longer period, like a couple of days or a week, this increase in energy consumption is likely to be minimal, mainly resulting from storage losses. This is because actions like charging your EV will already account for these needs in the following days.

3.2.3 Peak-Shedding

This concept aims to decrease the peak load demand, particularly when the demand puts transformers, feeder lines, or the grid itself at risk of reaching their thermal or supply limits. It is primarily achieved through direct load control (DLC) of household appliances, such as adjusting thermostat temperature setpoints for heaters, managing

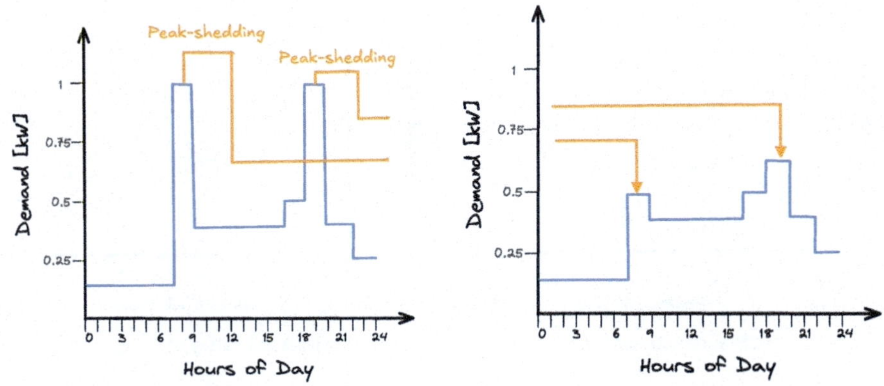

(a) Before peak-shedding. **(b)** After peak-shedding.

Fig. 3.5 The graph showcases the concept of peak-shedding or clipping. It involves the removal or complete elimination of load demand during on-peak periods, such as 7:00 AM to 9:00 AM and 6:00 PM to 8:00 PM, to assist the grid in maintaining its balance. Typically, this is implemented through direct load control (DLC) by grid operators. In Fig. 3.5a, the total energy consumption is 9.75 kWh, and it reduces to 8 kWh in Fig. 3.5b after shedding a load of 0.5 kW for 2 h and 0.375 kW for another 2 h

electric water heaters, and controlling air-conditioning units. Peak-shedding reduces the demand and consequently, the energy consumed by specific loads, which may occasionally result in a slight reduction in consumer comfort, particularly in extreme cases.

Figure 3.5 provides an example of the peak-shedding concept, which involves completely eliminating the peak demand of a household during a specific time and day. This can be achieved by actions such as stopping the operation of appliances like air conditioners or heat pumps, either by the consumers themselves or through demand-response programs such as DLC. Note that the application of peak-shedding leads to a decrease in energy consumption for that specific day. For a more comprehensive understanding of DLC and the specific hardware requirements, please refer to Sects. 3.3.2 and 3.4.4 respectively.

In summary, load-shifting aims to redistribute energy consumption to off-peak hours using various pricing and incentive-based schemes (see Sect. 3.3), while peak-shedding focuses on reducing energy usage during peak demand periods through measures like DLC to alleviate stress on the grid, prevent blackouts, and lower electricity costs for both consumers and utilities. Both strategies are essential components of demand-side management in optimizing energy usage and promoting grid stability.

3.3 Various Pricing and Incentive Schemes

In the preceding section, we discussed the key strategies of load-shifting, valley-filling, and peak-shedding in demand-side management. Understanding how these concepts are put into practice in real-world scenarios is crucial. Therefore, this section delves into the two primary implementation approaches of price- and incentive-based schemes. These approaches play a vital role in regulating the utilization of demand-side resources across various time scales, including real-time (milliseconds to less than 15 min), intra-day (one to several hours), day-ahead (one-day operational planning), month-ahead (monthly operational planning), and year-ahead (yearly system planning).

In descending order of timescale, on the one hand, the price-based schemes comprise time-of-use (ToU), day-ahead real-time pricing (RTP), and critical-peak pricing (CPP) as well as RTP. On the other hand, the incentive-based schemes include capacity and ancillary programs, demand bidding, interruptible and emergency programs, and direct load control (DLC). It is worthwhile to mention that in certain cases, both schemes can be implemented simultaneously to achieve the highest possible reduction in consumers' electricity usage.

For example, the grid operator (GO) can initiate day-ahead real-time pricing (RTP) and demand bidding at the day-ahead timescale. If these schemes prove ineffective, the GO can then implement RTP, critical-peak pricing (CPP), as well as interruptible and emergency programs. As a last resort, if necessary, the GO can implement direct load control (DLC), which grants control over consumer appliances based on the GO's requirements.

In Fig. 3.6, the different timescales for resources management in DSM are illustrated. The year-ahead timescale represents the longest time horizon, during which regulatory bodies and governments plan activities related to the grid system, such as expansion projects (e.g., adding transformers or installing new windmills) or activating/deactivating power plants. The month-ahead timescale focuses on operational activities that can be planned within a month's timeframe such as maintenance. The day-ahead timescale is used to plan operational activities one day in advance, such as forecasting energy prices or renewable energy generation for the following day. The intra-day timescale is utilized for planning operations within a specific day. Lastly, the real-time timescale involves activities or operations that need to occur within a very short period, typically not exceeding 15 min.

3.3.1 Price-Based Approaches

This scheme serves as an effective method to modify consumers' behavior regarding electricity usage. By adjusting electricity prices, it has the potential to encourage consumers to reduce consumption during periods of high prices and increase consumption during times of negative prices when there is an excess of energy available

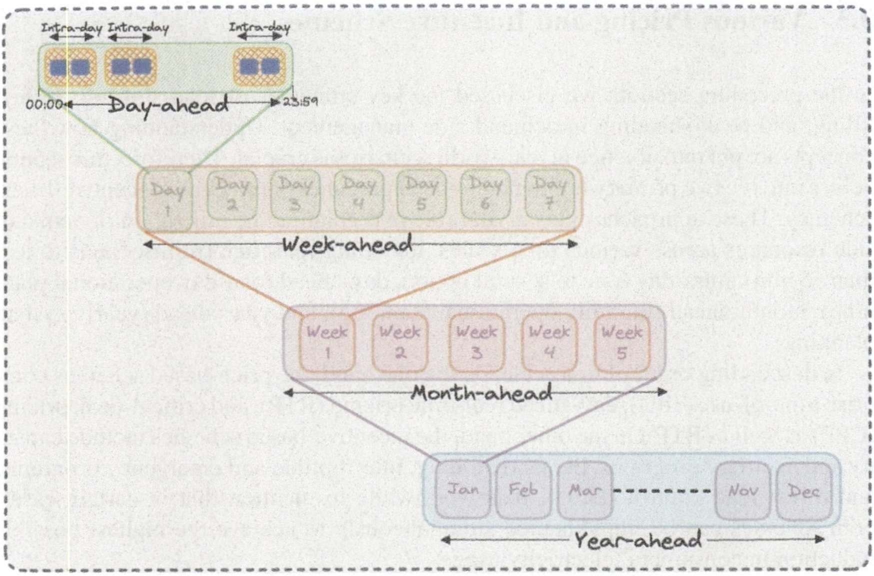

Fig. 3.6 Graphical presentation of different timescales considered in demand-side management (DSM). The smallest timescale (e.g., real-time), represented by the dark blue rectangle, ranges from milliseconds to less than 15 min. The orange rectangles depict the intra-day timescale, which occurs within a single day and spans from 1 to several hours. The green rectangles illustrate the day-ahead timescale, encompassing the entire day from 00:00 to 23:59. Additionally, DSM allows for defining actions within the week-ahead (red rectangles) and month-ahead (purple rectangles) timescales. The year-ahead timescale (light blue rectangle) is the longest period encompassing twelve months of the year. Please be aware that the example provided pertains to months that contain either 31 days or span across 5 weeks

from renewable sources. In the realm of pricing strategies, three effective schemes can be identified: time-of-use (ToU), real-time pricing (RTP), and critical-peak pricing (CPP).

The basic structure of a time-of-use (ToU) pricing scheme involves the establishment of two distinct prices: on-peak and off-peak, also referred to as daytime and nighttime tariffs. This pricing scheme falls within the timescale range depicted in Fig. 3.6, specifically between the year-ahead and month-ahead timescales. Typically, the on-peak or daytime period spans from 7:00 AM to 9:00 PM, while the remaining hours are categorized as off-peak or nighttime. It is important for the price difference between these two periods to be significant in order for the ToU pricing scheme to be effective.

It should be noted that the specific periods and price differentials are subject to contractual agreements and may vary between different utility companies or distribution system operators (DSOs) within the same region. Please be aware that the practical application of time-of-use (ToU) pricing in the real energy markets does not necessitate the use of advanced and costly Smart Meters. A traditional analog

Fig. 3.7 A photograph captured by the author displaying an analog meter with two counters labeled NT (NiederTarif in German meaning low tariff) and HT (HochTarif in German meaning high tariff), which represent off-peak and on-peak tariff consumption respectively. Any sensitive information present in the image has been obscured

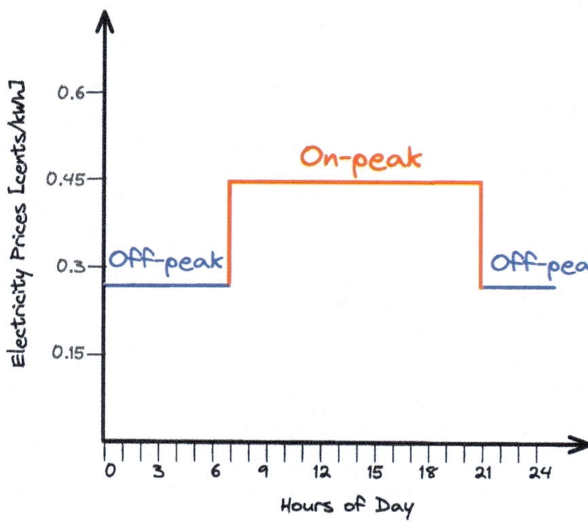

Fig. 3.8 Time-of-use or ToU scheme consists of in general two pricing levels namely off-peak (blue line) and on-peak (red line) tariffs. In the given example the electricity during on and off-peak times costs 45 and 28 cents/kWh respectively. It is important to note that the prices depicted in this figure are solely for illustrative purposes

meter equipped with two counters, one for on-peak consumption and another for off-peak consumption, is sufficient for calculating monthly energy costs as depicted in Fig. 3.7. The two pricing levels of ToU are illustrated in Fig. 3.8, with blue and red lines depicting off-peak and on-peak tariffs respectively.

The real-time pricing (RTP) scheme establishes prices on an hourly or half-hourly basis, reflecting the fluctuations in the intra-day or day-ahead costs of generating and delivering electricity. Typically, RTP offers two pricing options: "one-part" pricing, where all electricity consumption is charged at the hourly or spot price, and "two-part" pricing, which incorporates a historical baseline of consumers' electricity usage. In the two-part RTP approach, consumers are billed based on the hourly prices only for

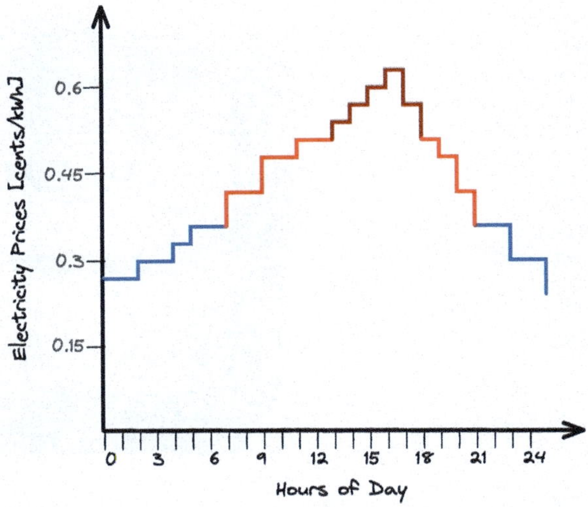

Fig. 3.9 Under the real-time pricing (RTP) scheme, the electricity cost varies hourly or half-hourly, mirroring the fluctuations in the wholesale electricity price. In this particular example of "one-part" RTP, the low electricity prices are represented by the blue line (ranging from 0.28 to 0.35 cents/kWh), the medium prices are depicted by the light red line (ranging from 0.36 to 0.5 cents/kWh), and the high prices are indicated by the dark red line (ranging from 0.51 to 0.65 cents/kWh). It is important to note that the prices depicted in this figure are solely for illustrative purposes

the additional electricity used above or below the baseline. This way, consumers are exposed to market prices only for the marginal usage.

Figure 3.9 illustrates an example of price fluctuations within the "one-part" real-time pricing (RTP) scheme for a specific day, showcasing the dynamic nature of electricity markets. It is important to acknowledge that implementing RTP in real-world scenarios presents challenges, as it necessitates the presence of Smart Meters at each consumer's location, along with reliable Internet connectivity. This can be particularly challenging in rural areas where access to such infrastructure may be limited.

The critical-peak pricing (CPP) scheme combines elements of both time-of-use (ToU) and real-time pricing (RTP) schemes. Under normal grid conditions, the CPP scheme operates similarly to ToU, with defined off-peak and on-peak tariffs. However, during system contingency times, the prices during on-peak periods experience a significant increase, similar to the RTP scheme. This adjustment is made to manage the sudden surge in demand.

It is important to note that CPP prices are limited to a specific number of hours per year, such as during major events like football matches, or extreme weather conditions (e.g., warm or cold). During these critical peak hours, electricity prices are significantly higher than standard rates to reflect the increased cost of generating and delivering electricity during times of high demand, and the exact timing of these events is unknown in advance.

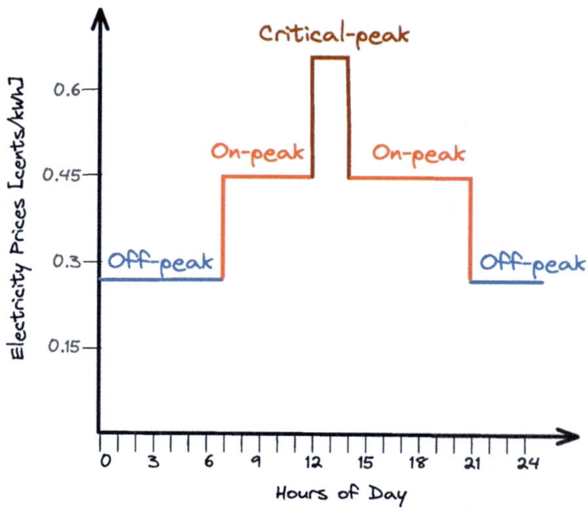

Fig. 3.10 Critical-peak pricing is a combination of time-of-use (ToU) and real-time pricing (RTP) schemes. Under normal grid conditions, the ToU scheme is in effect, represented by the blue line (off-peak tariff of 28 cents/kWh) and light red line (on-peak tariff of 45 cents/kWh). However, during critical situations, the on-peak prices experience a significant increase, similar to RTP, as shown by the dark red line (critical-peak tariff of 65 cents/kWh). It is important to note that the prices depicted in this figure are solely for illustrative purposes

The goal of critical-peak pricing is to incentivize customers to reduce or shift their electricity usage away from these peak periods, thereby helping to alleviate strain on the grid, reduce the need for expensive and polluting peaking power plants such as diesel generators, and lower overall electricity costs. By charging higher prices during critical peak hours, customers are encouraged to conserve energy, use energy-efficient appliances, and participate in demand response programs to lower their electricity bills.

Figure 3.10 provides an example of a CPP scheme, illustrating how, during grid issues (in the given example, between 12:00 PM and 2:00 PM), the prices under ToU increase significantly (65 cents/kWh instead of 45 cents for on-peak and 28 cents/kWh for off-peak periods). The critical-peak pricing (CPP) scheme operates at the intra-day timescale depicted in Fig. 3.6, allowing consumers a window of minutes or hours to take actions aimed at reducing their electricity usage.

3.3.2 *Incentive-Based Methods*

Like the price-based approaches mentioned earlier, incentive-based ones are highly effective in influencing consumers' electricity usage patterns. However, unlike the former, these approaches motivate consumers to decrease their consumption by offering incentives in the form of a percentage reduction in their monthly electricity bills. In the context of incentives, five effective methods can be identified: (1) capacity market programs, (2) ancillary services, (3) demand-side bidding, (4) emergency and interruptible programs, and (5) direct load control.

In capacity market programs, consumers contribute to the overall system's capacity by offering load curtailment or other demand-side resources as a means to replace or reduce the need for conventional generation methods by grid operators. Hence, the consumers agree and commit to reducing their electricity usage by a predetermined amount when system contingencies arise. These contingencies can include unexpected events such as equipment failures (e.g., transformers, feeder lines), extreme weather conditions (e.g., too hot or cold temperatures), or high-demand periods.

Consumers are typically notified in advance (e.g., month-ahead timescale) about these curtailment events, allowing them to plan accordingly. Incentives for participating in these programs often include financial rewards, reduced electricity rates, or other benefits such as upfront reservation payments (e.g., payments due to committing to perform curtailment when asked), while penalties may be imposed if consumers fail to meet their curtailment commitments.

Ancillary services are additional support services necessary for maintaining the reliability and stability of the power system. These services encompass activities such as regulating frequency, controlling voltage, and maintaining reserves. Demand-side management plays a vital role in providing ancillary services by utilizing the flexible demand-response capabilities of consumers' resources. This helps in balancing the grid and addressing fluctuations in electricity supply and demand. As part of this process, consumers have the opportunity to bid load curtailments in electricity markets in advance (e.g., month-ahead timescale) to serve as operational reserves.

If their bids are accepted, they are compensated at the market price for committing to be on standby and ready to curtail their electricity usage if needed. When their load curtailments are required, the respective grid operators will contact them, and they will be paid based on the spot market energy price. It is important to note that the specific timescales and operational details of capacity market programs and ancillary services can vary depending on the region, market structure, and regulatory framework in place.

In demand-side bidding, consumers have the opportunity to bid on specific load reductions in the wholesale electricity market. If their bid is lower than the prevailing market price, it is accepted. In such cases, consumers are obligated to curtail their electricity usage by the amount specified in their bid. Failure to do so may result in penalties. Typically, demand-side bidding programs attract participation from large consumers, such as data centers and industrial sites, with a capacity of one megawatt or more. However, even small consumers can participate in these programs with the assistance of third parties, such as aggregators.

Aggregators help by combining the loads of multiple small consumers and submitting bids on their behalf in the market. For instance, Entelios AG, which was acquired by EnerNOC in 2014, is a prominent provider of solutions for demand-response aggregators, including virtual power plants, in Europe. These aggregators play a crucial role in enabling small consumers to participate in demand-side bidding programs and contribute to load reductions in the electricity market (Marra et al. 2020). It is worthwhile to note that the demand-side bidding programs usually operate at the day-ahead timescale.

Within the intra-day timescale, the incentive-based scheme offers both emergency and interruptible/curtailable programs. The emergency program provides consumers with incentive payments for reducing their electricity usage when the system is experiencing a shortage of reserves. Consumers who participate in this program receive incentives for load reductions during emergency conditions. On the other hand, the interruptible/curtailable programs integrate load reduction options into retail tariffs and contracts. Participants in these programs agree to reduce their electricity usage during system contingencies and, in return, receive a rate discount or bill credit.

The specific reduction targets are predefined and participants are expected to meet these targets. Traditionally, interruptible programs have been available primarily to large industrial or commercial consumers. However, these programs may also be extended to other consumer segments depending on the program's terms and conditions. It is important to note that penalties may be imposed for failing to curtail as outlined in the program's terms and conditions. The severity of these penalties will vary depending on the specific program.

When all other methods of incentivizing and managing electricity demand have been exhausted, the direct load control (DLC) approach can be implemented as a last resort. This approach grants utilities and grid operators the authority to remotely control specific electrical appliances owned by consumers, including washing machines, refrigerators, air conditioners, and water heaters, among others. Under DLC, utilities can either turn off these appliances entirely or adjust their temperature setpoints. For example, they may lower (increase) the temperature of a water heater (air conditioner).

To facilitate this remote control, a bi-directional communication system is required. This system enables program operators to swiftly and remotely manage the electrical appliances with short notice. DLC programs primarily target residential or small commercial consumers and operate in real-time or near real-time timescales. They are designed to provide a means of managing electricity demand during critical periods when supply is strained. By temporarily controlling the operation of certain appliances, utilities can effectively reduce overall electricity consumption and alleviate stress on the grid.

It is important to note that DLC is typically implemented as a measure of last resort, when other incentive-based and price-based methods have been exhausted. Its purpose is to ensure the stability and reliability of the power system during times of high demand or system contingencies. Table 3.1 provides an overview by associating the various incentive and price-based programs mentioned earlier with the four timescales depicted in Fig. 3.6, which are adopted in demand-side management.

Table 3.1 A mapping of various incentive and price-based schemes to the different timescales adopted in demand-side management

Timescale	Schemes	
	Price	Incentive
Month-ahead	Time-of-use	Capacity market Ancillary services
Day-ahead	Real-time pricing	Demand bidding
Intra-day	Real-time pricing Critical-peak pricing	Emergency programs Interruptible programs Curtailable programs
Real-time		Direct load control

3.4 Components for Realization

Let's consider a scenario where the grid operator (GO) predicts a power shortage situation at 12:00 PM the next day using prediction models for generation, consumption, and electricity prices. The GO communicates with consumers through bi-directional communication channels and requests a demand-response (DR) event in day-ahead. Based on contractual agreements and consumer participation in such demand-side management (DSM) programs, the GO executes direct load control (DLC). During the DR event, the GO needs to continuously monitor the demand and consumption of each participating consumer by accessing their Smart Meters.

The existence of bi-directional communication enables the GO to gather real-time information. However, not all electronic appliances at the consumer's end may be "Smart" and equipped with communication interfaces for control purposes. To enable the GO to perform DLC, non-"Smart" appliances require hardware support. This allows the GO to remotely control non-"Smart" appliances such as turning them off (e.g., washing machine, water boiler) or adjusting their temperature setpoints (e.g., water boiler, air conditioner).

Thus, to successfully implement DSM in practical scenarios, four crucial components are required, as depicted in Fig. 3.11: prediction models, Smart Meters, bi-directional communication, and hardware support. Without any of these components, the implementation of DSM becomes unfeasible. This section gives a brief overview on those four components respectively.

3.4.1 *Prediction Models*

Generating prediction models for demand-side management (DSM) is a complex task that requires in-depth knowledge found in specialized literature. This section will provide a brief overview of forecasting, focusing only on the aspects relevant

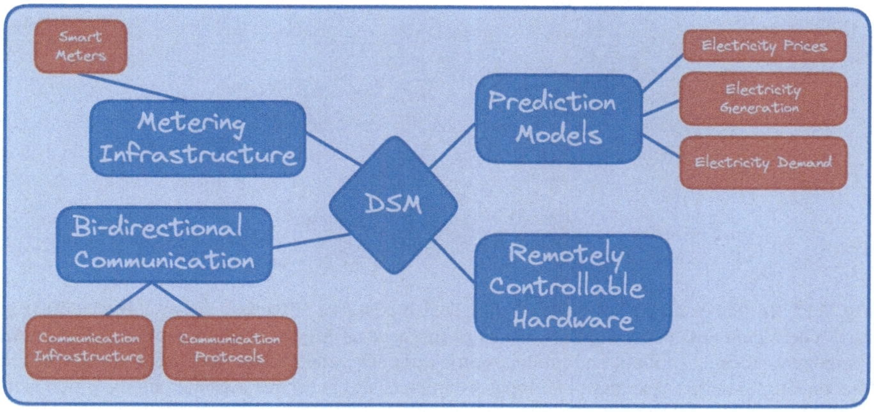

Fig. 3.11 The realization of demand-side management becomes possible with the presence of these four essential components of prediction models, remotely controllable hardware, metering infrastructure, and bi-directional communication

to generic concepts. Basically, there are two approaches to generating prediction models: analytical and data-driven.

The analytical approach utilizes laws of physics and mathematics to generate forecasts. For instance, using Joule's law (Joule 2011), the power demand (P) can be estimated by multiplying the voltage (V) by the current (I), as shown in the equation $P = V \cdot I$. These models require accurate input parameters (such as V and I in the given example) for precise predictions, which can sometimes be challenging or even require forecasting.

Analytical models mathematically describe the underlying system and are generally complex to derive, as they need to account for all relevant phenomena that influence the system's evolution over time. However, once the model is constructed, it offers high accuracy. Figure 3.12 illustrates the concept of the analytical modeling approach, where incoming and outgoing arrows represent input and output, and blue rectangles depict components that interact to represent specific system phenomena. Further details on this modeling approach, specific to ICT and data centers, will be discussed in Chap. 6.

Alternately, the data-driven approach relies on the presence of a dataset as input and generates forecasts as output (see Fig. 3.13). In the context of DSM, this dataset typically comprises time-series data. To put it simply, a time-series dataset organizes data points in a sequential manner based on time. The time interval between two data points can vary, ranging from milliseconds to seconds, minutes, hours, and so on.

Figure 3.14 illustrates a basic example of a time-series dataset that displays ambient temperature setpoints for a winter day, with intervals of 1 h. Typically, grid operators provide information with a more detailed granularity of 15-minute intervals. By leveraging a time-series dataset, it becomes possible to develop models that predict

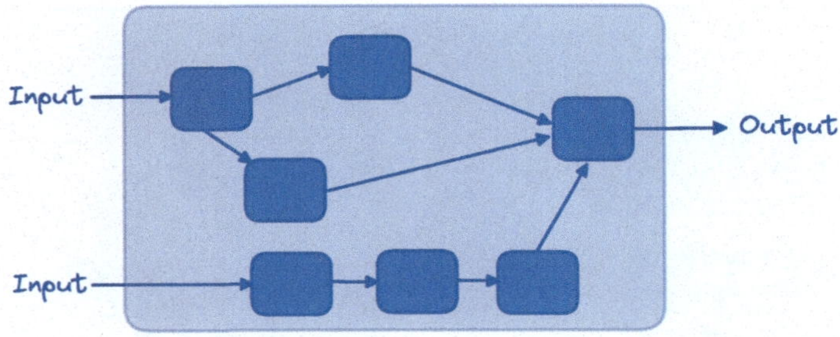

Fig. 3.12 In this example, we have an analytical model that comprises seven components (e.g., dark blue rectangles) capturing the various phenomena of a given system mathematically. Each component receives an input and produces an output. The underlying model (e.g., the large light blue rectangle) in this case takes two input parameters and generates a single output, which is the result of the interactions of the different components. Note that the output can be interpreted as one single prediction

Fig. 3.13 This is an illustration of creating a data-driven model that involves several inputs and produces an output by identifying correlations among the data points. Note that the model does not need to capture the underlying components of the given system and their interactions

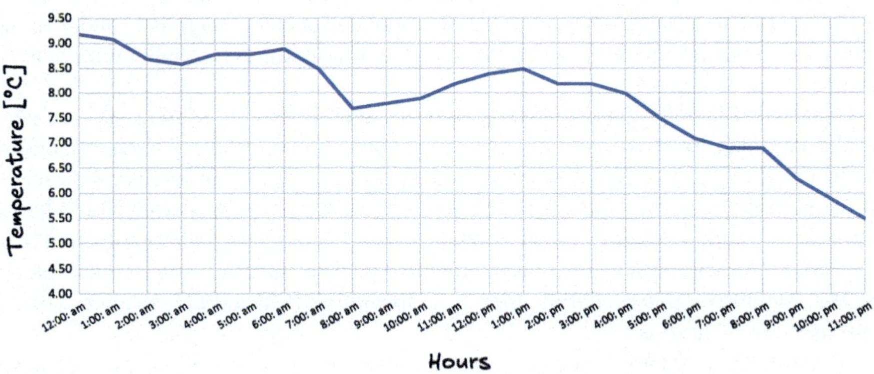

Fig. 3.14 A time-series representation of the ambient temperature for a winter day, graphically displayed with the Y-axis representing temperature values in Celsius and the X-axis indicating 1-hour intervals

three crucial aspects: (1) electricity generation from renewable energy sources, (2) electricity consumption and demand by consumers, and (3) electricity prices.

In data-driven approaches, we commonly encounter two methods: statistical and machine-learning. Both methods rely on historical data, but they differ in how they approach the analysis. Statistical methods establish functions by examining relationships between pairs of data points, such as the correlation between electricity generation and weather conditions. On the other hand, machine-learning methods automatically identify influential features from the historical data to make accurate forecasts. It's important to note that there is a wide range of data-driven models available in the literature, which can be categorized based on the forecasting window: short-term, medium-term, and long-term.

Short-term forecasting involves generating forecasts for the next day, while medium-term and long-term forecasting refer to predictions for the next week, month, or even year. Each of these categories can also be broken down into smaller forecasting intervals, such as 15 min, hourly basis, or larger intervals of 6 h. It's worth mentioning that the smaller the forecasting interval (e.g., 15-minute intervals), the more challenging it becomes to generate accurate predictions.

In the field of DSM and forecasting renewable energy generation as well as consumption of the consumers, auto-regression integrated moving average-based (ARIMA) methods such as SARIMAX, SARIMA, and ARIMAX are widely used statistical data-driven approaches (Siami-Namini et al. 2018). The "S" prefix in these methods indicates seasonality, while the "X" suffix represents the influence of exogenous variables on the forecasts. To ensure accurate predictions using ARIMA-based methods, it is crucial to have a dataset with minimal anomalies, such as missing values and outliers, as these can negatively impact the quality of the predictions.

To identify the exogenous variables in the dataset, Pearson's correlation analysis (Cohen et al. 2009) is commonly performed, comparing different independent variables with the dependent (e.g., the predicted) one. Variables showing a high positive correlation (e.g., a Pearson's correlation value greater than 0.75) are considered good candidates for serving as eXogenous variables in SARIMAX and ARIMAX methods.

In previous studies, researchers have proposed ARIMA-based models to forecast the percentage of renewable energy sources (Basmadjian and Shaafieyoun 2022, 2023; Basmadjian et al. 2021). Similarly, statistical models utilizing ARIMA have been developed for forecasting electricity load, with references available in Cho et al. (1995), Fard and Akbari-Zadeh (2014), Hor et al. (2006), Juberias et al. (1999), Lee and Ko (2011), Nie et al. (2012) and Tarmanini et al. (2023). For a comprehensive overview of these models, readers can refer to Alfares and Nazeeruddin (2002) and Singh et al. (2012). Additionally, ARIMA-based models have been suggested for predicting electricity prices by Conejo et al. (2005), Contreras et al. (2003), Jakaša et al. (2011), Zhao et al. (2017), Zhou et al. (2004, 2006).

Various machine-learning methods have been proposed in the literature to produce forecasts for renewable energy generation, energy consumption of consumers, and electricity prices over time. For a detailed overview of these methods, readers are encouraged to consult reference (Aslam et al. 2021). Recurrent neural networks (RNN), long short-term memory (LSTM), and gated recurrent units (GRU) are among the extensively studied methods in this field (Hewage et al. 2021; Liu et al. 2021; Mishra et al. 2020; Rajagukguk et al. 2020; Wang et al. 2017). Comparisons have shown that LSTM and GRU models tend to produce more accurate results compared to RNN. However, it is important to note that these two models are more complex and require longer training times than the RNN.

Data privacy and security concerns have also received significant attention in this area. Traditionally, the aforementioned time-series machine-learning methods (e.g., LSTM, GRU, RNN) were trained using a centralized learning (CL) scheme. In this approach, consumers such as homes would send their complete data to a central server hosted in a data center, which would then generate a single prediction model for all the consumers. To address privacy concerns, the concept of federated learning (FL) was introduced.

This approach eliminates the need for consumers to transmit their data to a central server. Instead, they generate a forecasting model using their local data and send this model to a central server. Unlike centralized learning, where the central server trains the model, the federated learning scheme aggregates models from multiple consumers into a single model that aims to accurately represent the diverse requirements of those consumers.

However, both learning schemes still rely on a central server for model training (in the case of centralized learning) and model aggregation (in the case of federated learning), which introduces the issue of a single point of failure (SPoF). If the central server fails, all computations are lost. To address the problem of SPoF in both centralized and federated learning schemes, a novel approach has been proposed that enables learning in a peer-to-peer (P2P) manner.

In this learning scheme, the logical arrangement of peers (e.g., graph or network structure) determines the learning process. Each peer, such as a consumer, is connected to neighboring peers to collaboratively train their respective machine-learning models. This P2P learning scheme eliminates the need for a central server for training (as in CL) or coordination (as in FL), effectively resolving the SPoF issue. The P2P learning scheme follows a two-stage process.

First, each peer trains a local model using its own data, known as local learning. Then, during the collaborative learning stage, peers exchange their locally generated models with their neighboring peers. This exchange allows each peer to receive different models from its peers, which aids in generating the final model. Importantly, since peers exchange only their local models and not their data, the P2P learning scheme also addresses privacy concerns present in CL and FL schemes.

Recent studies (Basmadjian et al. 2022; Boubouh et al. 2023a, b) have demonstrated that the P2P learning scheme not only resolves the SPoF issue and privacy concerns as mentioned above but also exhibits high energy-efficiency and accuracy comparable to the other two learning schemes. This is because the training process is executed on commodity devices such as mobile phones or IoT devices instead of running it on power-hungry servers in data centers.

Figure 3.15 visually represents the three learning schemes discussed earlier: CL, FL, and P2P learning. It emphasizes the necessity of a central server for CL and FL, while the P2P learning scheme operates without requiring one. Additionally, the figure demonstrates how each learning scheme generates machine-learning models. It is worth noting that various methods, such as LSTM, GRU, or RNN, among others can be employed for training in these three learning schemes.

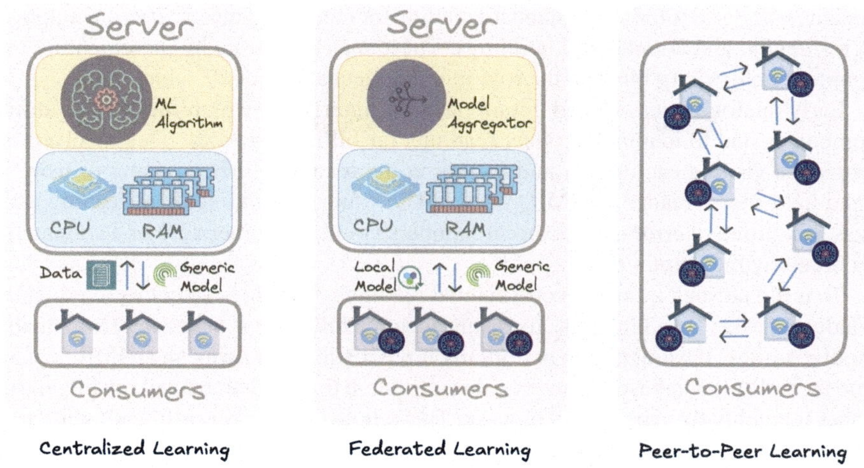

Fig. 3.15 Three distinct machine-learning schemes of centralized (CL), federated (FL), and peer-to-peer (P2P). In both CL and FL, there is a server involved, unlike in P2P learning. Consequently, the server becomes a single point of failure (SPoF), meaning that if it crashes, all machine-learning operations will be lost. Additionally, it is important to note that in CL, consumers send their data to the central server, which may raise concerns about data privacy. However, in FL and P2P learning, consumers exchange their models (not data) with the server and peers, respectively

There is a trade-off between model accuracy and training time, which directly impacts energy consumption. Configuring relevant hyperparameters, including the number of epochs, batch size, iterations, layers, and architecture, presents a challenging task. To address this challenge, blackbox methods like Bayesian optimization and model-free methods have been proposed (Wu et al. 2019).

Due to the computational demands of numerous machine-learning methods, pure blackbox optimization becomes prohibitively expensive. As a solution, multi-fidelity methods have been suggested, which utilize less computationally intensive versions of the blackbox function to approximate the evaluation of hyperparameter configurations (Feurer and Hutter 2019). For a more comprehensive exploration and review of this research area, readers are encouraged to refer to reference (Torres et al. 2021).

3.4.2 Smart Meter

Smart Meters play a crucial role in the implementation of demand-side management (DSM) within the Smart Grid. This capability is not achievable with traditional analog meters. To understand why, it is important to explore the historical evolution of the metering infrastructure.

Analog electricity meters, also referred to as electromechanical meters, have been utilized for many decades. They were initially introduced in the early 20th century and became widely adopted as the standard method for measuring electricity consumption in residential and commercial buildings. These meters operate by incrementing the counter of electricity through the rotation of a mechanical disc.

Early analog meters lacked communication interfaces, which meant that utility companies had to manually collect readings for billing purposes. This involved an employee visiting each meter and recording the reading either on paper or using a hand-held meter reader (HHMR) device. Unfortunately, this manual reading process was prone to errors, as incorrect numbers could be deliberately or accidentally recorded by the employee.

To mitigate such human errors, manual readings of analog meters were typically conducted every 2 or 3 months. To ensure accurate billing, it is necessary to manually input readings (if using paper) or digitally input readings (if using an HHMR device) into the utility's databases. However, it is important to note that manual reading introduces a potentially error-prone process. This whole process is costly and inefficient which results in delays from when the meter is read to when the bill is sent to the customer.

In 1977, Tesla introduced the concept of automatic meter reading (AMR) as an alternative to manual entry or physical inspections conducted by utility employees. This innovative approach involved equipping meters with a one-way radio-frequency (RF) communication interface, enabling remote transmission of meter readings. To collect the meter readings, a specialized piece of equipment installed in a drive-by vehicle is used. This equipment collects readings when both the meters and the equipment are within the RF range, eliminating the need for manual or HHMR device-based readings by utility employees.

The collected readings are then digitally synchronized with the utility company's database to generate bills. While this automated process improves efficiency and generates accurate energy bills, it is important to note that it does not support basic or advanced services of DSM. However, it significantly reduces errors and streamlines the meter reading process, benefiting both the utility company and the consumers.

With the introduction of Smart Meters, a more efficient and advanced form of metering has become possible. The basic Smart Meters are equipped with an interface allowing for bi-directional communication via the Internet between the utility company and the consumers. Thus, they enable utility companies to remotely and continuously access meter readings without the need for physical intervention by employees, as required in the previously mentioned cases. In addition to measuring energy consumption, these meters can also monitor other important key performance indicators (KPIs) such as demand, voltage, frequency, harmonics, and more.

These basic Smart Meters also have elementary control features to enable some DSM programs. Furthermore, some advanced Smart Meters (ASM) have two interfaces: one for external communication via the Internet, and another for internal communication within the consumer's premises usually wireless via Wi-Fi. Those meters offer the capability for utility companies to remotely control home appliances thus implementing DSM programs such as DLC (see Sect. 3.3.2). Also, ASM can promptly report power outages back to the utility, facilitating faster service restoration.

Table 3.2 Evolution of the meter reading process from communication and infrastructure perspectives

Communication level	Metering infrastructure			
	Analog meter	Automatic meter reading	Smart meter	Advanced smart meter
No communication	Manual meter reading through utility employee			
Low		One-way RF meter reading by drive-by equipment		
Medium			Internet-based meter reading of KPIs and control functions to support some DSM programs	
High				Internet-based meter reading of KPIs and advanced DSM programs like DLC, outage reporting, etc.

This digital metering technology, being basic or advanced Smart Meters, allows consumers to monitor their energy consumption more accurately, empowering them to make informed choices about their energy usage. By enabling anytime and frequent meter readings, they not only enhance the efficiency and accuracy of billing processes but also aim to reduce operational costs while providing consumers with precise and timely information about their energy consumption.

Table 3.2 provides an overview of the development of the metering infrastructure, highlighting the communication requirements associated with each stage. It demonstrates that analog meters do not have any communication capabilities, while advanced Smart Meters require a high level of communication to support advanced DSM programs such as DLC, power outage reporting, etc. The primary function of the Smart Meter on the consumers' side is to accurately and continuously acquire system-related parameters and send those back to the utility company as illustrated in Fig. 3.16.

This involves first measuring the magnitude and frequency of the voltage, as well as the magnitude and phase displacement of the current relative to the voltage. These parameters are used to calculate KPIs such as power factor, active/reactive power, and total harmonic distortion (THD). To measure the current, a current-sensing shunt

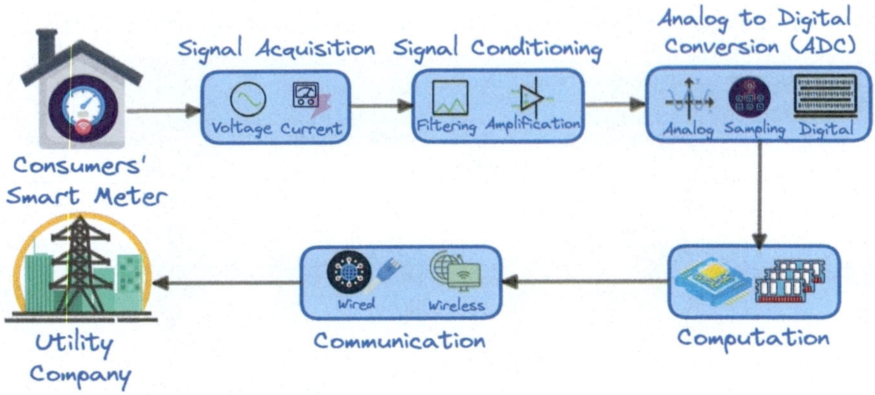

Fig. 3.16 The five different stages of a Smart Meter for monitoring power-relevant KPIs and transferring them to the utility company

resistor is used at the input channel, typically located at the load of the consumers' premise. For voltage measurements, a resistive voltage divider is used at the input channel, usually at the point of supply. Once the voltage and current signals are acquired continuously, they undergo a conditioning stage to prepare them for conversion (e.g., ADC stage). This may involve filtering, amplification, attenuation, offset addition, or subtraction.

Since continuous analog signals cannot be used for computation or transmitted through digital communication mediums, the corresponding signals need to be converted from analog to digital at the ADC stage. It is crucial to accurately reconstruct the continuous signal from its discrete digital samples without any loss of information. To prevent inaccuracies caused by aliasing, the samples are selected based on the Nyquist-Shannon sampling theorem (Landau 1967), which states that the sampling rate must be at least twice the highest frequency of the continuous signal. By adhering to the Nyquist sampling rate, the resulting digital sample points can faithfully reproduce all frequencies of the original continuous analog signal at the receiver's end (e.g., utility company).

Once the sampling process is complete, the next step involves converting the sampled points into a digital format for further processing and transmission. This conversion is achieved through the use of quantization techniques, where the number of levels is determined by the chosen number of bits. In the case of a specific KPI, such as voltage, a 3-bit analog-to-digital converter (ADC) can represent 8 (i.e., 2^3) different levels within a given range.

Figure 3.17 visually depicts this quantization example, for a range of 1.6 V, spanning from −0.8 V to 0.8 V with increments of 0.2 V. Moving from left to right, the sample points (represented by red dots) are converted into binary format based on the level they fall into. In this specific example, the binary representation of the sample points is as follows: 001, 011, 100, 101, 110, 111, 110, 101, 011, 010, 001, 000, 001,

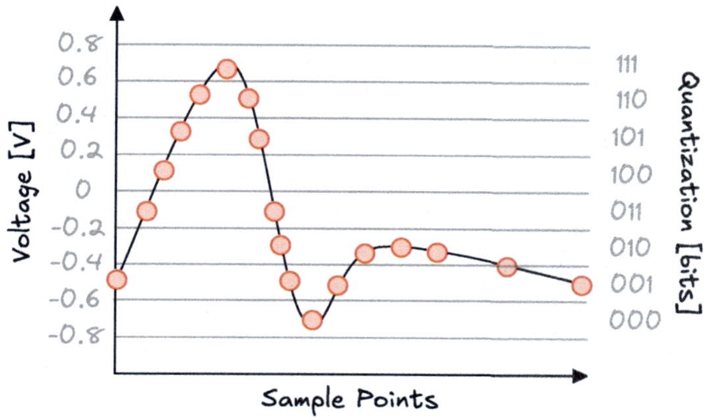

Fig. 3.17 In this example, we observe a quantization process applied to a sampled function represented by red dots. The quantization is performed using a 3-bit level, which allows for the representation of voltage variations between −0.8 V and +0.8 V in increments of 0.2 V. This is achieved by dividing the voltage range of 1.6 V into 2^3 (or 8) equal steps

010, 010, 010, 010, 001. This results in a total of 18 sample points, transforming the corresponding continuous signal into a digital one.

After the analog signals have been converted into digital format, they can undergo additional processing and transmission, as both of these processes require binary data. The computation stage is responsible for performing various operations on the input signals, such as arithmetic calculations, timestamping, data preparation for communication or output devices, handling of routines related to irregular inputs (e.g., payment, tamper detection), data storage, and system updates, among others.

To facilitate these tasks, Smart Meters are equipped with a processor capable of parallel execution and memory modules for storing and buffering data. Smart Meters utilize various network adapters for communication, offering both wired and wireless options. Wired options include public switched telephone network (PSTN), power line carrier (PLC), cable modems, and Ethernet. Wireless options include Wi-Fi, Zigbee, Infrared, and GSM/GPRS/CDMA cellular networks.

3.4.3 Bi-directional End-to-End Communication

To facilitate demand-side management (DSM) programs, a communication infrastructure is established with three interfaces: the home area network (HAN), the neighborhood area network (NAN) or last-mile network, and the wide area network (WAN), as shown in Fig. 3.18.

The HAN plays a crucial role in providing connectivity between various home appliances (e.g., washing machines, refrigerators, televisions, ovens), heating and

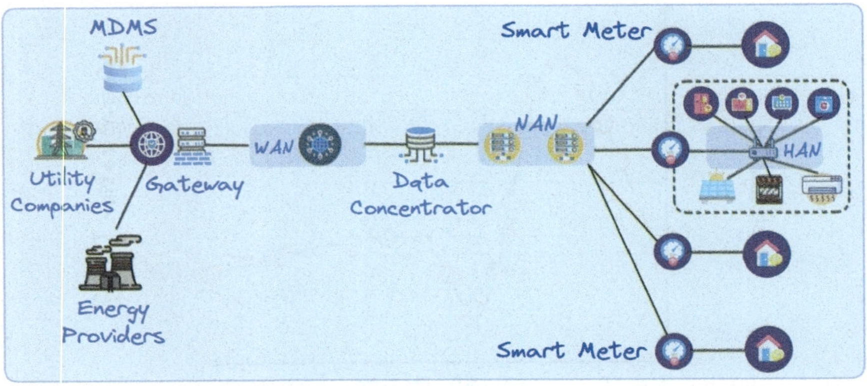

Fig. 3.18 To effectively implement demand-side management programs, it is essential to have a comprehensive infrastructure that includes three interfaces: home area network (HAN), neighborhood area network (NAN), and wide area network (WAN). These interfaces facilitate bi-directional end-to-end communication between consumers and energy providers, utility companies as well as the providers of meter data management system (MDMS)

air conditioning (HVAC) systems, and renewable energy sources like solar panels. This connectivity is typically established through a wireless medium, such as a Wi-Fi router, which is also linked to the Smart Meter. As a result, the Smart Meter can collect energy-related data from all the connected appliances and components within the HAN for monitoring purposes.

Additionally, it can forward commands, such as turning off or adjusting temperature setpoints, from the grid operator (GO) to the appliances, in order to implement demand-response programs like DLC. It is important to note that standardized wired or wireless protocols, such as Wi-Fi, are used in the HAN to ensure interoperability. Moreover, the HAN facilitates the implementation of a home energy management system (HEMS), which centrally manages the energy consumption within a specific consumer (e.g., household).

Through the HAN, the HEMS can perform the following functions: (1) monitor the energy usage of various appliances and present the information on a dashboard, (2) control the heating, ventilation, and air conditioning (HVAC) system, and (3) manage smart appliances and smart plugs. Additionally, the HEMS incorporates robust security measures to protect consumer data and the metering system. As a result, both the consumer and the GO can confidently carry out remote operations on home appliances. The implementation of the HEMS can be achieved through either (1) a cloud-based service, (2) a dedicated IoT-based controller, or (3) advanced Smart Meters.

The objective of the neighborhood area network (NAN), also known as the last-mile area network, is to establish connectivity between the Smart Meters of different consumers and the data concentrator. This enables the transfer of consumption readings from the Smart Meters to the data concentrator. Additionally, the NAN facilitates the transmission of diagnostic messages, firmware updates, and real-time or near real-time messages (such as prices) to support the power system. The specific communication technology employed in the NAN for DSM applications may vary depending on the implementation.

Commonly used technologies in Smart Grid systems for NAN include power line carrier (PLC), wireless technologies like Wi-Fi or Zigbee, and cellular networks such as 4G or 5G. The choice of communication technology depends on factors such as coverage, data transfer requirements, reliability, and cost-effectiveness within the DSM deployment. For example, if Zigbee technology is employed, which offers a data transfer rate of 250 kb/s, each consumer would only utilize the communication link for a fraction of a second per day to transmit their energy consumption information to the data concentrator. This efficient utilization of the communication link ensures that the overall system operates smoothly and effectively.

The wide area network (WAN) plays a crucial role in enabling end-to-end communication between GOs and consumers in the context of Smart Grid and DSM. It serves as the final link in the chain, facilitating the relay of data concentrators with third parties such as suppliers. These third parties utilize the metering data for various purposes like billing, planning, and management. To establish this communication, a diverse range of wired and wireless technologies are employed. Wired options include Ethernet-based communication protocols like TCP/IP, which are renowned for their reliability and ability to transmit data at high speeds.

Additionally, fiber optic cables are utilized to ensure fast and dependable communication between data concentrators and WANs. These cables are capable of efficiently transmitting large volumes of data over long distances. On the wireless front, cellular technologies such as 3G, 4G, and 5G are extensively used. These cellular networks offer broad coverage and possess the capacity to handle substantial amounts of data. Another wireless option is WiMAX (Worldwide Interoperability for Microwave Access), which provides long-range connectivity for data concentrators to connect with WANs.

In scenarios where infrastructure is limited or in remote areas, satellite-based communication proves to be highly convenient. Satellite communication enables the establishment of connectivity between data concentrators and WANs, ensuring seamless communication even in challenging environments. By leveraging this wide array of wired and wireless technologies, the WAN acts as a vital bridge, facilitating efficient and effective communication between data concentrators and third parties within the Smart Grid and DSM ecosystem.

The meter data management system (MDMS) is a software application (e.g., a database) or platform that is responsible for collecting, storing, managing, and analyzing the data generated by Smart Meter of the consumers. It serves as a central repository for all the metering data collected from various Smart Meters deployed in the grid. It has the following six primary features:

- Data collection: It collects data from Smart Meters, which can include energy consumption data, interval data, demand-response signals, and other relevant information.

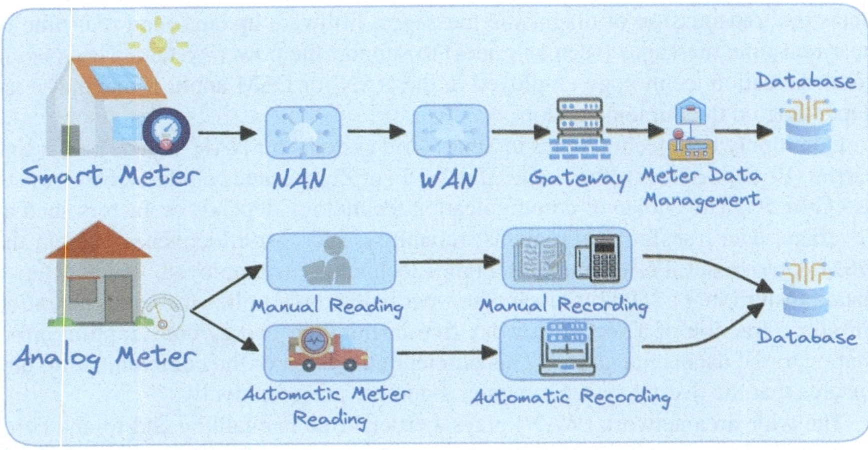

Fig. 3.19 The process of collecting and storing meter readings into the meter reading's database both from traditional analog meters and Smart Meters

- Data storage: The collected data is stored securely in the MDMS database, ensuring its integrity and availability for further analysis and processing.
- Data validation and estimation: It performs validation checks on the collected data to identify any anomalies or errors. It may also estimate missing or incomplete data based on predefined algorithms.
- Data management: The MDMS manages the metering data throughout its lifecycle, including data retrieval, archiving, and retention policies.
- Data analytics: It provides tools and functionalities to analyze the metering data, enabling utilities to gain insights into energy consumption patterns, identify trends, and make informed decisions for grid optimization and demand management.
- Integration: The MDMS integrates with other systems in the Smart Grid ecosystem, such as billing systems, customer information systems, and distribution management systems, to enable seamless data exchange and interoperability.

Thus, the MDMS plays a crucial role in efficiently managing and utilizing the vast amount of data generated by Smart Meters, enabling utilities to improve grid operations, enhance customer services, and optimize energy management in the Smart Grid and DSM ecosystem.

Figure 3.19 depicts the process of collecting and storing meter readings in the database of MDMS. In a Smart Grid environment, Smart Meters, along with HANs, NANs, and WANs, enable the transmission of energy-related key performance indicators (KPIs) such as power demand, voltage, current, and energy consumption to the MDMS automatically without human intervention. This information is then stored in the underlying database.

In contrast, traditional analog meters require manual input by humans for reading and storing information, making automation of the whole process almost impossible.

For manual meter readings, an employee from the utility company either records the readings on paper or uses a digital hand-held device called HHMR (see Sect. 3.4.2 for more details). In the case of automatic meter reading (AMR), the meter readings obtained through one-way RF communication channels are automatically stored in a computing system or device.

The final step involves synchronizing the meter readings with the MDMS database. The process of recording readings on paper is prone to errors and is time-consuming, as it requires manual insertion of the information into the database. However, for HHMR and AMR, the synchronization process is relatively straight-forward and requires minimal human intervention. MDMS often utilizes database technologies that are capable of handling large volumes of meter data efficiently. Some commonly used database technologies for MDMS include:

- Relational database management systems (RDBMS) such as Oracle, MySQL, Microsoft SQL Server, and PostgreSQL, are commonly used for storing and managing meter data. These databases provide structured data storage and support SQL queries for data retrieval and analysis.
- Time-series databases like InfluxDB, Prometheus, and TimescaleDB are designed specifically for handling time-stamped data, making them suitable for storing and analyzing meter data that is collected at regular intervals.
- NoSQL databases like MongoDB, Cassandra, and Apache HBase are used in some MDMS implementations. These databases offer high scalability and flexibility, making them suitable for handling large volumes of meter data with varying data structures.
- Data warehouses technologies like Amazon Redshift, Google BigQuery, and Snowflake are used for storing and analyzing meter data in a structured and optimized manner. These databases are designed for handling large-scale data analytics and reporting.

In general, the progress in technologies, particularly in terms of end-to-end communication infrastructure and databases as mentioned previously, has greatly simplified the collection and storage of meter readings. This has resulted in a reduction of errors and manual labor involved in managing data. It also paved the way for performing advanced analytics as well as predictions on energy consumption. Next, we'll explore the final component of the DSM framework: the hardware support necessary for effectively managing and controlling remotely the equipment involved in this process.

3.4.4 Remotely Controllable Hardware

It is not mandatory for every power-demanding home appliance to have remotely controllable interfaces, such as smart washing machines or smart TVs, for the implementation of demand-side management (DSM). However, to enable remote control

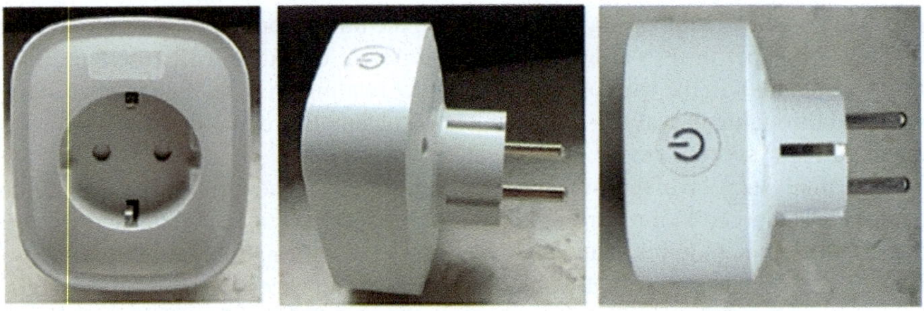

Fig. 3.20 A simple example of a load control switch (LCS), which is a device offering assistance for loads that do not have the capability for remote control through the Internet. This enables the grid operator to remotely activate or deactivate the connected loads

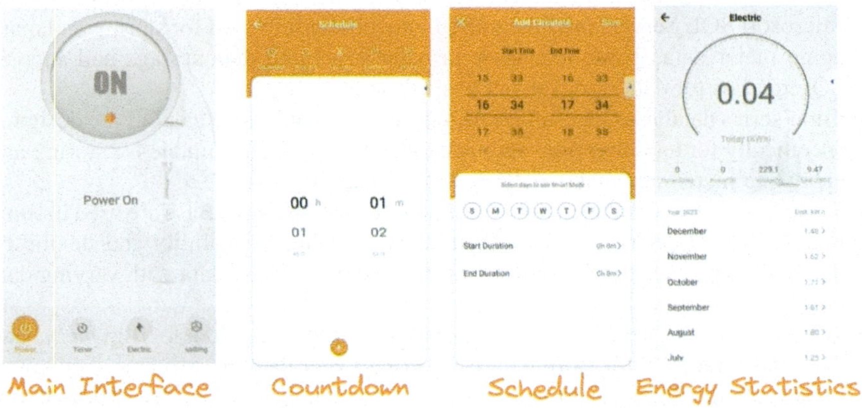

Fig. 3.21 The dashboard of the web application controlling the socket-like LCS of Fig. 3.20

of such appliances for tasks like demand-response (DR), it is necessary to have hardware support in the form of load control switches (LCSs), controllable or smart thermostats (ST), and adjustable speed drives (ASD).

The load control switch (LCS) is an electronic device comprising two main components. Firstly, it includes a communication module that connects with the home area network (HAN) to receive control signals from the grid operator. Secondly, it incorporates a relay that can activate or deactivate the load connected to the respective LCS. It is important to note that the duration for which the appliance remains disconnected can typically be adjusted using a built-in clock.

Figure 3.20 presents an example of a basic socket-like LCS that is equipped with Wi-Fi capability to activate or deactivate the connected load. This device can be controlled remotely, as demonstrated by the main interface shown in Fig. 3.21. For instance, users can turn the load ON or OFF by pressing the "Power" button. The "Timer" button allows for scheduling ON or OFF events by specifying a specific

Fig. 3.22 A basic type of controllable or smart thermostat that enables consumers to conserve energy for heating purposes by remotely adjusting the temperature setpoints

schedule or using a countdown format. Additionally, the device provides the ability to gather statistics on energy consumption for the current and previous months (historical data), current demand, voltage, and other relevant information.

Controllable or smart thermostats (ST) offer an advanced alternative to traditional programmable thermostats, allowing consumers to efficiently manage their energy usage in residential heating and air conditioning systems (HVAC) through remote control and configuration. Similar to LCS, these controllable thermostats are equipped with a communication module that enables external connectivity, typically utilizing Wi-Fi or Zigbee protocols. This enables remote control and configuration of room temperature setpoints. In terms of heating, the primary function of these smart thermostats is to control the opening or closing of water valves within the connected heating system.

Figure 3.22 illustrates an example of a basic controllable thermostat designed for heating purposes. It showcases how these thermostats can be installed as replacements for traditional programmable thermostats. The main interface of the web application for this controllable thermostat is depicted in Fig. 3.23. It presents various features for controlling heating temperature setpoints. These features include the ability to select different operation modes such as "Auto" (which automatically opens and closes the valve to meet the consumer's configuration), "Manual" (which functions

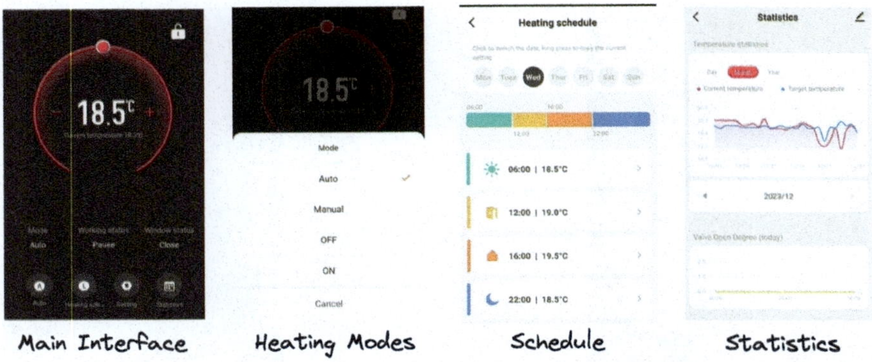

Fig. 3.23 The dashboard of the web application controlling remotely the temperature setpoints of a smart thermostat

like a traditional programmable thermostat), "ON", and "OFF" (which turns off the entire heating system of the smart thermostat).

The thermostat offers scheduling options with different operating temperatures and provides statistics in daily, monthly, and yearly views to compare the configured (e.g., wished) and actual temperature setpoints. When it comes to controlling air conditioning (AC) units, controllable thermostats can be utilized. These thermostats are equipped with either Bluetooth or infrared connectivity to establish a connection with the AC. Additionally, they can be controlled remotely by consumers using either Wi-Fi or Zigbee interfaces.

Figure 3.24 provides a visual representation of a basic controllable thermostat that utilizes Wi-Fi and infrared technologies for communication. It is crucial to understand that both LCS and ST devices require a connection to the home area network (HAN) via the router when utilizing the Wi-Fi protocol. However, when using the Zigbee protocol, an additional Zigbee-compatible hub is required, which is then connected to the router of HAN either through wired (e.g., Ethernet) or wireless (e.g., Wi-Fi) communication means.

Pumps, ventilation units, compressors, and similar equipment often utilize electric motors, where the power demand and energy consumption are closely tied to the motor's speed or rotations per minute (RPM). To account for worst-case scenarios, manufacturers typically configure the electric motor to operate at its highest speed or frequency, even during situations of medium to low load. However, the majority of motorized appliances experience varying loads over time and are frequently operated below full capacity.

Adjustable speed drives (ASDs) offer a solution by allowing motors to meet the required operational conditions while optimizing power and energy usage when the system is not operating at maximum load. Many directly connected motors for pumps and fans are often oversized, and replacing them with ASDs can result in significant energy savings. A similar concept of ASD is applied to processors in

Fig. 3.24 An illustration of a simple thermostat device with Wi-Fi and infrared control capabilities is an example of a device that can connect to an air conditioning (AC) unit using infrared signals and receive control signals from consumers through the Wi-Fi interface

electronic appliances like computers, mobile phones, and IoT devices, known as dynamic voltage and frequency scaling (DVFS) technology.

With DVFS, processors are equipped with a range of frequencies, similar to RPM, spanning from the lowest to the highest. During high-workload situations, the CPU is configured to run at the highest frequency to deliver optimal performance. However, during medium to low workload situations, instead of operating the CPU at the highest frequency and wasting unnecessary energy, DVFS technology allows for configuration at lower frequencies, enabling energy savings to be achieved. Chapter 4 will provide a more in-depth exploration of the DVFS technology used in processors. It will delve into the intricacies of DVFS and offer a comprehensive analysis of the trade-off between energy consumption and performance when implementing DVFS in processors.

3.5 Summary

The increasing global energy consumption resulting from modern society's reliance on electricity for daily activities has rendered the traditional "supply follows demand" paradigm impractical and inefficient. As a viable alternative, the new paradigm of "demand-side management" (DSM) has emerged, offering easier integration of intermittent renewable energy sources.

This chapter extensively explored the benefits and potential of DSM within the context of Smart Grid, presenting a mutually beneficial scenario for consumers and utility companies. Furthermore, the chapter introduced three practical concepts adopted in DSM: shifting (e.g., rescheduling) and shedding (e.g., removing) the demand, along with valley-filling, which involves increasing energy consumption during off-peak periods.

We then delved into the various timescales employed in the operations of DSM, ranging from yearly periods to millisecond intervals. Numerous pricing (e.g., time-of-use, real-time pricing, critical-peak pricing) and incentive-based (e.g., capacity markets, ancillary services, demand bidding, emergency programs, direct load control) schemes are explained and exemplified based on these timescales.

Lastly, the chapter outlined the essential components required for implementing DSM in the context of Smart Grid, emphasizing the need for four pillars of (1) prediction models, (2) Smart Meters, (3) a bidirectional communication infrastructure, and (4) remotely controllable hardware support. In terms of prediction models, this chapter provided a brief overview of both analytical (e.g., whitebox) and data-driven (e.g., blackbox) techniques, discussing the strengths and weaknesses of each.

Furthermore, the chapter showcased various studies in the literature that utilize regression-based ARIMA and machine-learning approaches to generate forecasts related to electricity demand, renewable generation, and prices. Additionally, we demonstrated the potential of peer-to-peer machine learning as a promising scheme, offering improved energy efficiency compared to traditional centralized and federated learning schemes. This approach also addresses concerns such as the single point of failure and data privacy issues associated with those traditional schemes.

In the final part, we delved into the progression of communication requirements for meter readings, specifically focusing on the transition from analog meters with no communication capabilities to advanced Smart Meters with high communication needs. We also provided an overview of the various stages involved in collecting meter readings using Smart Meters, illustrating the quantization process as an example.

The section on bidirectional end-to-end communication explored the essential interfaces, such as the home area network (HAN), neighborhood area network (NAN), and wide area network (WAN), that facilitate communication between consumers and utility companies.

Additionally, the chapter introduced the necessary database technologies for implementing the meter data management system (MDMS). In the end, the chapter touched upon remotely controllable hardware, including load control switches, smart and controllable thermostats, and adjustable speed drives. Real-life examples were provided to illustrate their practical applications and controlling capabilities.

References

Alfares HK, Nazeeruddin M (2002) Electric load forecasting: literature survey and classification of methods. Int J Syst Sci 33(1):23–34

Aslam S, Herodotou H, Mohsin SM, Javaid N, Ashraf N, Aslam S (2021) A survey on deep learning methods for power load and renewable energy forecasting in smart microgrids. Renew Sustain Energy Rev 144:110992

Basmadjian R, Boubouh K, Boussetta A, Guerraoui R, Maurer A (2022) On the advantages of p2p ml on mobile devices. In: Proceedings of the thirteenth ACM international conference on future energy systems, e-Energy '22, pp 338–353, New York, NY, USA. Association for Computing Machinery

Basmadjian R, Niedermeier F, de Meer H (2016) Analysis of incentive-based demand response mechanism. https://tu-dresden.de/bu/wirtschaft/bwl/ee2/ressourcen/dateien/dateien/ordner_enerday/enerday-2016/Basmadjian.pdf, April 2016. Accessed 02 Dec 2024

Basmadjian R, Shaafieyoun A (2022) Arima-based forecasts for the share of renewable energy sources: the case study of Germany. In: 2022 3rd international conference on smart grid and renewable energy (SGRE), pp 1–6

Basmadjian R, Shaafieyoun A (2023) Assessing arima-based forecasts for the percentage of renewables in Germany: insights and lessons for the future. Energies 16(16)

Basmadjian R, Shaafieyoun A, Julka S (2021) Day-ahead forecasting of the percentage of renewables based on time-series statistical methods. Energies 14(21)

Boubouh K, Basmadjian R, Ardakanian O, Maurer A, Guerraoui R (2023a) Efficient and accurate peer-to-peer training of machine learning based home thermal models. In: Proceedings of the 14th ACM international conference on future energy systems, e-Energy '23, page 524-529, New York, NY, USA. Association for Computing Machinery

Boubouh K, Basmadjian R, Ardakanian O, Maurer A, Guerraoui R (2023b) Peptm: an efficient and accurate personalized p2p learning algorithm for home thermal modeling. Energies 16(18)

Cho M, Hwang J, Chen C (1995) Customer short term load forecasting by using arima transfer function model. In: Proceedings 1995 international conference on energy management and power delivery EMPD'95, vol 1, pp 317–322. IEEE

Cohen I, Huang Y, Chen J, Benesty J, Benesty J, Chen J, Huang Y, Cohen I (2009) Pearson correlation coefficient. In: Noise reduction in speech processing, pp 1–4

Conejo AJ, Plazas MA, Espinola R, Molina AB (2005) Day-ahead electricity price forecasting using the wavelet transform and Arima models. IEEE Trans Power Syst 20(2):1035–1042

Contreras J, Espinola R, Nogales FJ, Conejo AJ (2003) Arima models to predict next-day electricity prices. IEEE Trans Power Syst 18(3):1014–1020

Fard AK, Akbari-Zadeh M-R (2014) A hybrid method based on wavelet, Ann and Arima model for short-term load forecasting. J Exper & Theor Artif Intell 26(2):167–182

Feurer M, Hutter F (2019) Hyperparameter optimization. In: Automated machine learning: methods, systems, challenges, pp 3–33

Hewage P, Trovati M, Pereira E, Behera A (2021) Deep learning-based effective fine-grained weather forecasting model. Pattern Anal Appl 24(1):343–366

Hor C-L, Watson SJ, Majithia S (2006) Daily load forecasting and maximum demand estimation using Arima and Garch. In: 2006 international conference on probabilistic methods applied to power systems, pp 1–6. IEEE

Jakaša T, Andročec I, Sprčić P (2011) Electricity price forecasting-arima model approach. In: 2011 8th international conference on the European energy market (EEM), pp 222–225. IEEE

Joule JP (2011) The scientific papers of James Prescott joule, vol 1. Cambridge University Press

Juberias G, Yunta R, Moreno JG, Mendivil C (1999) A new Arima model for hourly load forecasting. In: 1999 IEEE transmission and distribution conference (Cat. No. 99CH36333), vol 1, pp 314–319. IEEE

Landau H (1967) Sampling, data transmission, and the nyquist rate. Proc IEEE 55(10):1701–1706

Lee C-M, Ko C-N (2011) Short-term load forecasting using lifting scheme and Arima models. Expert Systems with Applications 38(5):5902–5911
Liu X, Lin Z, Feng Z (2021) Short-term offshore wind speed forecast by seasonal Arima-a comparison against gru and lstm. Energy 227:120492
Marra A, Carlei V, Baldassari C (2020) Exploring networks of proximity for partner selection, firms' collaboration and knowledge exchange. the case of clean-tech industry. Business Strategy Environ 29(3):1034–1044
Mishra S, Bordin C, Taharaguchi K, Palu I (2020) Comparison of deep learning models for multivariate prediction of time series wind power generation and temperature. Energy Reports 6:273–286
Nie H, Liu G, Liu X, Wang Y (2012) Hybrid of Arima and svms for short-term load forecasting. Energy Procedia 16:1455–1460
Rajagukguk RA, Ramadhan RA, Lee H-J (2020) A review on deep learning models for forecasting time series data of solar irradiance and photovoltaic power. Energies 13(24):6623
Shann M, Alan A, Seuken S, Costanza E, Ramchurn SD (2017) Save money or feel cozy?: a field experiment evaluation of a smart thermostat that learns heating preferences. In: Proceedings of the 16th conference on autonomous agents and multiAgent systems, vol 16. International Foundation for Autonomous Agents and Multiagent Systems (IFAAMAS)
Siami-Namini S, Tavakoli N, Namin AS (2018) A comparison of arima and lstm in forecasting time series. In: 2018 17th IEEE international conference on machine learning and applications (ICMLA), pp 1394–1401. IEEE
Singh AK, Khatoon S, Muazzam M, Chaturvedi D, et al (2012) Load forecasting techniques and methodologies: a review. In: 2012 2nd international conference on power, control and embedded systems, pp 1–10. IEEE
Tarmanini C, Sarma N, Gezegin C, Ozgonenel O (2023) Short term load forecasting based on Arima and Ann approaches. Energy Reports 9:550–557
Torres JF, Hadjout D, Sebaa A, Martínez-Álvarez F, Troncoso A (2021) Deep learning for time series forecasting: a survey. Big Data 9(1):3–21
Wang Y, Long M, Wang J, Gao Z, Yu PS (2017) Predrnn: recurrent neural networks for predictive learning using spatiotemporal lstms. In: Advances in neural information processing systems, 30
Wu J, Chen X-Y, Zhang H, Xiong L-D, Lei H, Deng S-H (2019) Hyperparameter optimization for machine learning models based on Bayesian optimization. J Electr Sci Technol 17(1):26–40
Zhao Z, Wang C, Nokleby M, Miller CJ (2017) Improving short-term electricity price forecasting using day-ahead lmp with Arima models. In: 2017 IEEE power & energy society general meeting, pp 1–5. IEEE
Zhou M, Yan Z, Ni Y, Li G (2004) An Arima approach to forecasting electricity price with accuracy improvement by predicted errors. In: IEEE power engineering society general meeting, 2004, pp 233–238. IEEE
Zhou M, Yan Z, Ni Y, Li G, Nie Y (2006) Electricity price forecasting with confidence-interval estimation through an extended Arima approach. IEE Proc-Gener Trans Distrib 153(2):187–195

Part II
Exploring and Exploiting Flexibilities in Data Centers

The second part of this book delves into the opportunities for flexibility and demand-side management (DSM) in data centers, focusing on their implementation through an energy management system (EMS). Chapter 4 explores the potential of DSM in data centers and the mechanisms that enable flexibility. It begins by emphasizing the importance of ICT and data centers in our digital systems, detailing key resources like servers, storage devices, networking equipment, cooling systems, and backup power systems. The chapter then discusses the feasibility of energy and demand management in data centers, introducing the concept of power usage effectiveness (PUE) to assess efficiency. It also examines demand flexibility in terms of time and amount flexibility, highlighting how workload, cooling systems, and UPS can provide flexibility. While cooling system and UPS mechanisms have no impact on service level agreements (SLAs), workload flexibility mechanisms can affect performance and SLAs, necessitating more relaxed SLAs for such services and DSM.

Chapter 5 discusses the importance of a feedback control loop in human-centered cyber-physical systems (HCPS), specifically focusing on data centers to optimize energy efficiency and demand-side management. It outlines a comprehensive process that starts with evaluating the HCPS environment, involving hardware, operators, and stakeholders, and ends with implementing optimal decisions through human-machine interfaces (HMI). The chapter then introduces a software-based orchestration of the energy management system (EMS), detailing components like data acquisition, analysis, and automation. It emphasizes an agent-based approach, highlighting how autonomous agents interact through allocation strategies and protocols, specifically the contract net protocol (CNP). Next, the chapter covers an EMS aimed at reducing energy consumption while meeting service level agreements (SLAs) for IT customers. Key components such as the "Demand Predictor" and "Optimizer" are discussed, showcasing the system's agnosticism to diverse hardware and software via connectors. A generic XSD template for describing ICT resources in data centers is also provided, serving as input for the optimization engine. Finally, the chapter addresses demand adaptation within an ecosystem comprising grid operators, data center operators, and IT customers. It critiques existing rigid SLAs and proposes relaxed SLAs based on operational context, supported by examples. Flexible service-level agreements (SDAs) and effective reward and penalty schemes are suggested, demonstrating that even with limited collaboration, data center operators have incentives to engage in demand-response activities. Overall, this section underscores the need for flexible agreements to facilitate effective demand-side management practices.

Chapter 4
DSM Opportunities and Demand Flexibility

4.1 Recapping the Current State of the Power Grid

In the past, human society functioned without relying on electrical power. However, in the modern world, the power grid has become essential and indispensable. This is due to the increasing demand for electricity caused by population growth and the emergence of energy-intensive systems like data centers. Additionally, the way electricity is generated and consumed is rapidly changing. These changes, as shown in this chapter, involve the integration of distributed but intermittent energy sources such as solar panels and windmills, as well as the strain on aging power grids.

As a result, traditional roles in the energy market are undergoing significant transformation. Today's consumers are not just passive users; they are becoming active participants (1) by generating their own power through renewable sources and (2) by offering energy storage solutions, such as home batteries. In contemporary energy markets, energy is increasingly regarded as a commodity, characterized by fluctuating prices that reflect real-time supply and demand. Consumers can now market their surplus energy back to the grid, creating a more decentralized and dynamic marketplace.

To comprehend the transformation of the power grid, it is crucial to identify the key factors driving this change. Traditionally, the responsibility of maintaining a balance between power supply and demand rested with the generation side. To achieve this equilibrium, different types of power plants were constantly adjusted to meet the demand, following a "supply follows demand" paradigm. As mentioned in Sect. 2.1, these power plants can be categorized into three main classes: base-load, medium-load, and peak-load.

Base-load power plants are utilized to meet the steady power demand in the grid. Power generation from base-load power plants, such as nuclear and large coal power plants, is cost-effective. However, their adaptability to sudden changes in power demand is limited, making them unsuitable for coping with rapid fluctuations. On the

R. Basmadjian, *Demand-Side Energy and Power Management in Data Centers*,
Lecture Notes in Energy 1, https://doi.org/10.1007/978-3-031-85956-4_4

other hand, peak-load power plants offer greater flexibility in adjusting their power output. However, they incur higher fuel costs and often have negative environmental impacts in terms of CO_2 emissions. Typical examples of peak-load power plants include natural gas or diesel generators.

The traditional form of the power grid is now revealing its limitations as new requirements arise, particularly from the integration of large amounts of renewable energy sources like solar panels and windmills. The unpredictable nature of these sources necessitates significant changes to the electricity infrastructure to ensure the future stability of the power grid.

Key changes include implementing more detailed monitoring of the grid and increasing the automation of grid management through the use of information and communication technologies (ICT). These requirements have led to the development of the Smart Grid, which combines ICT with power grid technology to enhance efficiency and reliability. To delve deeper into the role of ICT in Smart Grid, readers may consult Sect. 2.3 for additional information.

The emergence of the Smart Grid has brought attention to two important characteristics. Firstly, the intermittent nature of renewable energy sources introduces uncertainty and limited control over power generation from the supply side. This necessitates accurate forecasting of generation and consumption, as well as the need for flexibility on the demand side. Secondly, effective power generation management on the supply side and the utilization of flexibilities, such as load management, on the demand side require regular exchange of information in both directions.

ICT has been playing a significant role in enabling intelligent energy management for the Smart Grid. In this regard, demand-response (DR) schemes and demand-side management (DSM) have been proposed to leverage the flexibilities on the demand side (e.g., consumers). Specifically, DSM involves the interaction between the supply and demand sides, allowing consumers to adjust their power usage patterns and levels in response to generation scarcity (for more details refer to Sect. 3.2). DSM encompasses various schemes (for more details refer to Sect. 3.3) aimed at incentivizing the participation of the demand side in energy management.

4.2 Embracing the Digital Era

In the past few years, the integration of digital technology into our daily lives has become increasingly prominent, impacting various domains including business and personal social activities. Examples of this digital transformation can be seen in our online purchasing and payment methods, the convenience of online banking, and the widespread use of streaming services for entertainment purposes. The COVID-19 pandemic further accelerated our reliance on digital systems, as virtual platforms became essential for conducting conferences, lectures, and other activities. The digitized system has played a crucial role in facilitating our daily routines during these challenging times.

This widespread digitization has necessitated the development of a ubiquitous ICT infrastructure, which includes massive data centers operated by industry giants such as Google, Microsoft, Amazon, and Yahoo, to name a few. A data center serves as a centralized hub of networked computing systems, primarily utilized by organizations for remote storage, processing, and distribution of vast volumes of data. These networked computing systems rely on ICT resources, including servers, storage devices, and networking equipment such as routers and switches, to enable their functionality as shown in Fig. 4.1.

In simple terms, a data center can be likened to a large-scale personal computer (PC) housed within a building. Its main purpose is to store vast amounts of data, offering nearly unlimited storage capacity compared to regular PCs. Moreover, data

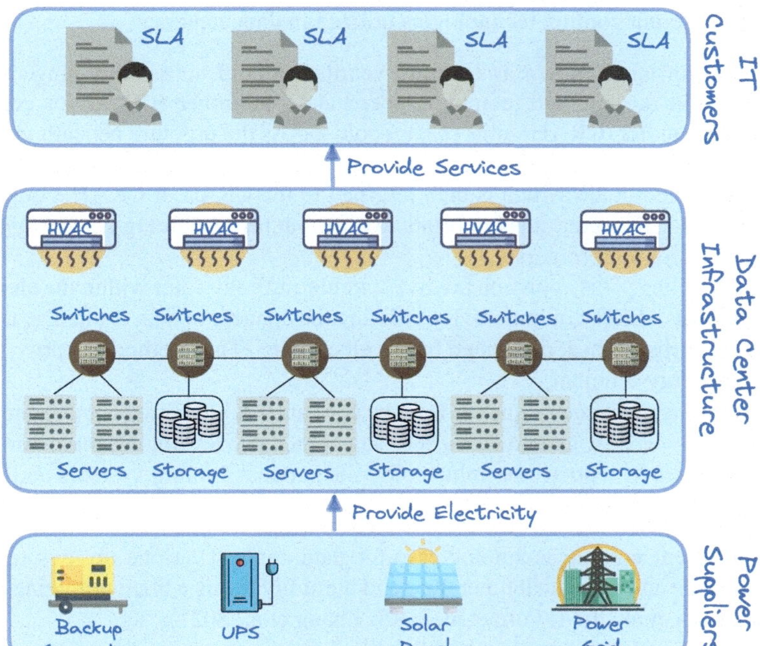

Fig. 4.1 A simplified overview of the components and entities involved in data centers can be presented as follows. From an infrastructure standpoint, the key ICT resources comprise servers, storage devices, and networking equipment like routers and switches. Cooling systems, such as HVAC, are employed to dissipate the heat generated by these ICT resources, particularly servers during computational tasks. The primary electricity supply for the ICT resources is sourced from the power grid. Some data centers also incorporate renewable energy sources like solar panels, which can partially or fully meet the power requirements of the ICT resources. Additionally, all data centers are equipped with alternative power sources like UPS and diesel generators, which are intended to be utilized in the event of power disruptions from the grid. Data centers offer services to their IT customers, with the quality of these services defined by contracts known as service level agreements (SLAs)

centers excel in processing information at an accelerated rate, thanks to their millions of CPU cores, as opposed to the limited processing power of a single multi-core processor in a PC.

Data centers prioritize uninterrupted operation. To achieve this, they incorporate redundant power supplies, including backup generators and uninterruptible power supply (UPS) systems. These measures ensure continuous functionality even during power outages. Those centers can be owned and operated by organizations for their internal use, or they can be third-party establishments that offer hosting and colocation services to businesses and individuals.

Unlike PCs that rely on fans for cooling, data centers employ advanced cooling systems to maintain optimal temperature and humidity levels. This is essential for protecting electronics of sensitive components such as servers, networking equipment, and storage devices from potential damage caused by excessive heat. There are three different cooling technologies utilized in data centers:

- Air cooling, also known as heating, ventilation, and air conditioning (HVAC) systems: In this method, computer room air conditioner (CRAC) or computer room air handler (CRAH) units release cold air. As the pressure beneath the raised floor increases, the cold air is directed into the equipment inlets. The cold air displaces the hot air, which is then returned to the CRAC or CRAH, cooled, and recirculated. However, it is important to note that this technology is highly energy inefficient for data centers.
- Liquid cooling: This approach involves circulating cold water within the electronic components of ICT equipment. By utilizing the concept of heat exchange, the cold water absorbs the heat dissipated by the electronics. This method has proven to be highly energy-efficient.
- Free cooling: This technique utilizes the cold air from the outside environment to cool down the heat generated by the data center. By leveraging the natural cold air, this method reduces the reliance on mechanical cooling systems and can be more energy-efficient.

These different cooling technologies offer data centers various options to effectively manage and regulate temperature and humidity levels while optimizing energy efficiency (Li et al. 2014; Pore et al. 2015; Zhang et al. 2021).

Figure 4.2 depicts the primary building blocks of data centers, which are comprised of enclosures and racks. Enclosures are similar to cupboards made of metal, as shown in Fig. 4.2a, and contain shelves or racks for the installation of ICT equipment such as servers and storage devices. The concept of enclosures offers a modular and flexible architecture that allows for the expansion of equipment. For example, additional servers and storage devices can be added to an existing enclosure if there are available slots, as illustrated in Fig. 4.2b.

Alternatively, a new enclosure can be integrated into the data center, enabling the installation of new equipment that can then be connected to the overall system. Figure 4.3a delineates an enclosure with all of its shelves occupied by servers and

(a) A set of almost empty enclosures. (b) Enclosures with installed servers and storage devices.

Fig. 4.2 This graphical representation showcases the fundamental components of data centers, which are enclosures. These enclosures serve as hosts for servers and storage devices, resembling shelves. Furthermore, there is interconnectivity between servers and storage devices within an enclosure and between different enclosures

storage devices. Additionally, Fig. 4.3b illustrates the interconnection between the equipment within the enclosure. It is important to note that this interconnection plays a crucial role in enabling the data center to provide various services and function as a unified system.

4.3 Chances of Demand and Energy Management

Due to the increasing demand for outsourcing support services and the growing communication needs, data centers have become increasingly important in the field of ICT. They are currently the fastest-growing sub-sector within ICT (Webb 2008).

(a) An enclosure with a set of servers and storage devices.

(b) Interconnectivity between the servers and storage devices of a given enclosure.

Fig. 4.3 This illustration demonstrates an enclosure that contains installed servers and storage devices, showcasing their interconnectivity

However, a significant drawback of this extensive computing activity is its substantial energy consumption (Katz 2009). It is widely acknowledged that the energy usage of the global ICT sector is comparable to that of the aviation industry, and this consumption continues to rise.

Consequently, the ICT sector contributes significantly to global CO_2 emissions. For example, in 2007, data centers in EU-27 consumed about 56 TWh of energy, in 2015 they consumed about 78 TWh in EU-28 (equivalent to 2.5% of total EU electricity use), and this figure is projected to exceed 160 TWh per year by 2030 (Koronen et al. 2020). Similarly, in 2014, data centers in the U.S. consumed an estimated 70 TWh, accounting for approximately 1.8% of the total electricity consumption in the country (Shehabi et al. 2016).

Currently, digital technologies, including data centers, contribute to approximately 5 to 9% of global electricity consumption. This percentage is expected to rise significantly as the world continues to embrace digitization and the proliferation of advanced technologies. Emerging innovations such as artificial intelligence (AI), the Internet of Things (IoT), blockchain, quantum simulations, and beyond 5G (B5G) networks are driving an unprecedented demand for computational power and data processing capabilities.

Energy management in a data center encompasses various considerations. The initial step involves selecting an appropriate location for construction, along with choosing a primary energy source and cooling system. Ideally, locations with cold climates like Finland are preferred, as they allow for the utilization of outside cold air (e.g., free cooling technology) to cool the heat generated by the data center. Alternatively, locations abundant in renewable energy sources enable the operation of the data center using green energy.

However, energy management extends beyond location selection to encompass infrastructure, equipment, and operational practices. Fortunately, even after the data center is constructed and equipped, there are opportunities during the operational phase to operate in an energy-efficient manner. For example, periodic replacement of energy-intensive servers with energy-efficient alternatives, such as ARM-based processors, which have gained popularity recently, is a viable option.

Additionally, data centers often over-provision ICT resources to ensure compliance with service level agreements (SLAs) of their customers. To address this, intelligent energy management systems (EMS) can be implemented. These systems focus on making optimal and energy-efficient decisions during the operation phase of the data center based on the prevailing circumstances. This book emphasizes the implementation and benefits of such intelligent EMS for data centers.

In the literature, various metrics have been proposed to evaluate and assess the energy efficiency of data centers. Among these, the power usage effectiveness (PUE) metric stands out as the most prominent (Horner and Azevedo 2016). PUE is calculated as the ratio between the total power demanded by the data center (P_{DC}) and the power needed by the ICT resources (P_{ICT}), represented by:

$$PUE = \frac{P_{DC}}{P_{ICT}} \tag{4.1}$$

A PUE value of 1 represents the highest level of energy efficiency in a data center, where all the power demanded and energy consumed is effectively utilized by the ICT resources for computation, storage, and communication purposes. On the other hand, inefficient data centers typically have PUE values equal to or greater than 1.5, indicating that 33% or more of the energy is wasted in performing these tasks.

To improve the energy efficiency of a data center and reduce its PUE value, it is crucial to explore optimization methods. By referring to Eq. (4.1), the power required by a data center can be broken down into three components: the power needed by the ICT resources (P_{ICT}), the power demanded by the cooling system (P_{CS}), and the power drawn by the lighting system (P_{LS}):

$$P_{DC} = P_{ICT} + P_{CS} + P_{LS} \tag{4.2}$$

Therefore, any EMS implemented in data centers aims to reduce the demand and subsequently the energy consumed by the ICT resources, which will indirectly lead to a reduction in energy consumption by the cooling system as well. It is worthwhile to note that energy (in Joules or Wh) is calculated by multiplying power (in Watts) with time (hours).

As a side note, in the realm of high-performance computing (HPC), the Green500 community currently relies on a standard metric for energy efficiency that measures the ratio of peak performance in GFlop/s to power demand in Watts during the execution of the high-performance Linpack (HPL) benchmark. This metric highlights that neither on-site renewable energy sources nor energy management systems (EMS) are accounted for; it focuses solely on the use of energy-efficient hardware. Therefore, further research is essential to integrate both renewable energy and EMS into the energy efficiency metrics for supercomputers and HPC systems.

Based on Google, through intelligent energy management, especially for cooling purposes, it was possible to reduce the average PUE of their data centers from 1.21 in 2008 to 1.1 in 2021 till 2023 (Google 2024). According to the literature (Basmadjian et al. 2016; Ghatikar et al. 2009, 2012), data centers are considered ideal candidates for participating in demand-response schemes and demand-side management for the following reasons:

- Data centers have highly automated and monitored ICT infrastructure, allowing them to take advantage of flexibilities without the need for human intervention.
- In certain cases, such as cloud computing, workloads in data centers are highly flexible, making it possible to curtail a portion of the total load within minutes.
- Data centers are often over-provisioned, meaning their ICT infrastructure is under-utilized most of the time. This presents opportunities for additional flexibilities.
- Data centers have a significant power demand, with some reaching up to 2.5 GW in Northern Virginia (see Chap. 8). Table 4.1 provides a list of the world's top 10 most power-demanding data centers in 2024 (Visual-Capitalist 2024).

Table 4.1 Power demand in MW of the world's top ten data centers based on Visual-Capitalist (2024)

Rank	Data center market	Country	Capacity (MW)
1	Northern Virginia	United States	2552
2	Beijing	China	1799
3	London	United Kingdom	1053
4	Singapore	Singapore	876
5	Tokyo	Japan	865
6	Frankfurt	Germany	864
7	Shanghai	Chine	725
8	Sydney	Australia	667
9	Dallas	United States	654
10	Silicon Valley	United States	615

Hence, the effective implementation of an appropriate EMS enables data centers to optimize their energy usage. The primary purpose of such a software system is to support data center operators (DCOs) in achieving their objectives in various aspects. This book focuses on two specific aspects: (1) minimizing overall energy consumption for contributing to energy efficiency, or (2) reducing peak power demand for demand-response and demand-side management programs.

Both aspects are appealing to DCOs as they aim to reduce operational expenditure (Opex), which increasingly constitutes a significant portion of a data center's total cost of ownership (TCO) (Barroso et al. 2019). The concepts and methodologies presented in this book are applicable to all types of data centers, including traditional computing (TC), high-performance computing (HPC), and cloud computing (CC). However, the extent of power and energy savings achieved will depend on the available flexibility of the workload and its corresponding SLAs, which can be different from one type to another.

4.4 Flexibility in Data Centers: Unlocking Potential

4.4.1 What Is Flexibility?

Demand flexibility refers to the ability to adjust planned activities by either preponing (e.g., shifting) or postponing (e.g., shedding) some or all of the demanded load from consuming entities, such as ICT equipment (e.g., servers) and cooling systems in data centers. This flexibility can be categorized into two dimensions: time-flexibility and amount-flexibility. Time-flexibility allows for the entire workload to be shifted or even shed, while amount-flexibility refers to the ability to decrease or increase the quantity of demand, typically measured in units of power, within a range from minimum to maximum.

Data centers, as significant consumers of energy in the power grid, have the potential to contribute to grid stability by adjusting their power demand as well as energy consumption when the grid is under-supplied. This requires data centers to possess flexibility, which refers to their ability to reliably modify their power demand to a desired extent and duration in response to changes in the residual load.

Flexibility is particularly crucial in future energy systems due to the integration of large-scale unpredictable renewable energy sources. To leverage this flexibility, DSM and demand-response schemes (Commission et al. 2008), such as OpenADR (Ghatikar and Bienert 2011), have become the standard approach. These schemes involve defining a set of actions for consumers, including data centers, to significantly reduce their electrical load during periods of power shortage or outages.

Quantifying and effectively managing the flexibility of data centers in terms of demand-response is essential to benefit both data center operators (by minimizing energy costs) and power grid operators (by ensuring grid stability). To gain a deeper understanding of the advantages and potentials of DSM from both consumers' and utilities' perspectives, we recommend referring to Sect. 3.1 and Fig. 3.2 for more detailed information.

4.4.2 *Exploring Mechanisms for Providing Flexibility*

Within data centers, various mechanisms exist that offer flexibility and can be leveraged for demand-response and demand-side management objectives. It is important to highlight that while certain mechanisms have no impact on service level agreements (SLAs) upon implementation, others may result in a potential degradation of quality of service (QoS). In such cases, additional agreements, such as relaxed SLAs (see Sect. 5.4.2), are necessary to ensure a balance between satisfactory performance and demand reduction.

4.4.2.1 Workload

The utilization of the underlying ICT infrastructure is directly related to the amount of work that needs to be completed by the corresponding services. SLAs ensure specific QoS characteristics for a given workload. In high-performance computing (HPC) data centers, where maximizing performance is a priority, the infrastructure is typically kept more than 75% utilized, allowing only a small window of time-flexibility without violating SLAs. Efficient rescheduling of jobs can lead to energy savings in this context (Mämmelä et al. 2012).

In contrast, typical cloud computing data centers offer some level of time-flexibility within SLAs (Quan et al. 2012). For example, backup services specify the frequency (e.g., daily) and availability (e.g., 99.99%) of backup operations, but do not specify the exact time (e.g., night or noon) for these operations. To manage workloads effectively, four mechanisms are described, which can be classified into two groups:

- Amount-flexibility: consolidation and frequency scaling.
- Time-flexibility: shifting and migration.

Next, we will begin by examining the amount-flexibility associated with consolidation and frequency scaling. Following that, we will introduce the other two mechanisms of time-flexibility namely shifting and migration.

Consolidation

This mechanism leverages the virtualization technology of ICT infrastructure to optimize resource utilization. By consolidating multiple workloads encapsulated within virtual machines (VMs) onto a subset of the data center's ICT equipment (e.g., servers), the number of active physical resources can be reduced. This is beneficial because the power demand of typical ICT resources is highly non-linear in relation to their utilization. For example, an idle server may still consume up to 50% of its maximum power demand (Basmadjian et al. 2012).

While this mechanism significantly reduces the overall power demand of the ICT equipment, it can also impact the performance of running services, particularly in dense consolidation scenarios where the ratio of workload to resources is high (e.g., 4:1 or more). The time it takes to react and implement consolidation depends on the configuration of the data center and the size of the workload, typically taking minutes (e.g., 1–3 min). Details on virtualization technology are given later in the "Migration" part.

In Fig. 4.4, an example is presented featuring two applications: a database running on Server 1 and data replication executing on Server 2. Both servers have a utilization rate of 30%. Assuming that Server 2 is more powerful and energy-efficient than Server 1, the concept of workload consolidation involves transferring the workload

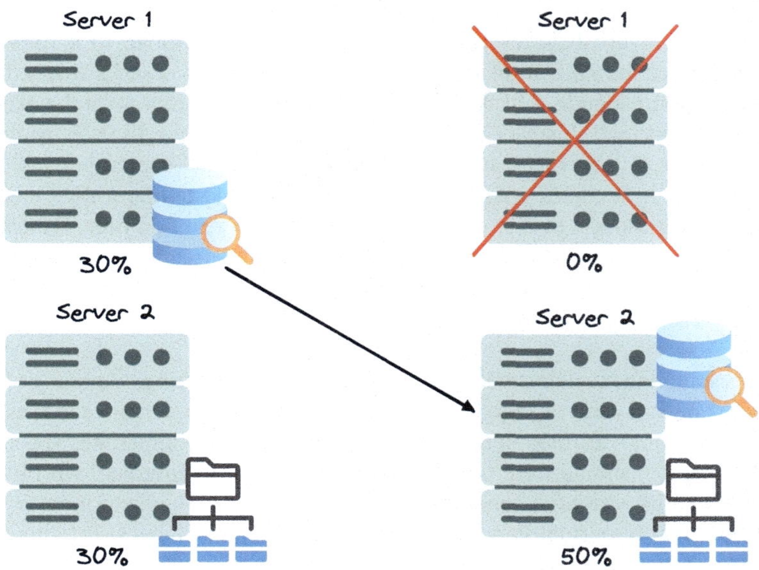

Fig. 4.4 On the left side, Server 1 is running a database application with a utilization rate of 30%, while Server 2 is running a data replication application with a utilization rate of 30% as well. On the right side, the database application from Server 1 is migrated to Server 2 and consolidated with the data replication application. As a result, Server 1 becomes idle and can be powered off to save energy, while Server 2 now has a utilization rate of 50%. Please note that the provided numbers related to the utilization rate are for illustrative purposes

associated with the database application from Server 1 and consolidating it with the workload of the data replication application on Server 2. As a result, Server 1 becomes idle and can be powered off to save energy if it is not required. Meanwhile, Server 2 experiences a utilization rate of 50%, resulting in a slight increase in energy consumption compared to the pre-consolidation scenario.

Frequency Scaling

This mechanism is related to hardware manufacturers and involves dynamically adjusting or scaling the frequency and voltage of the processor (known as DVFS) during periods of low to medium workload utilization rates (Snowdon et al. 2005; Suleiman et al. 2005). Major hardware vendors like Intel and AMD have developed their own implementations of DVFS, called SpeedStep and CoolnQuiet, respectively. From an operating system perspective, this mechanism is implemented through power policies or governors.

More details on the DVFS technology will be provided in Sect. 7.1. To understand the trade-off between performance degradation and power savings, analytical models like Markov chains proposed in Chap. 7 are necessary. Unlike consolidation, frequency scaling mechanism has an instantaneous reaction time and is typically executed through the governors (e.g., Linux's onDemand, Powersave, Performance, etc.) of the corresponding operating system.

Shifting

This mechanism involves rescheduling or postponing workload to a future time period to reduce the utilization of ICT resources and save power. However, for this to work, the corresponding SLA should not have strict constraints on the execution time of the workload. Instead, as long as the SLA guarantees 99.99% availability, it may not specify the exact timing of the backup operation (e.g., at noon or night). In such cases, it becomes possible to shift the workload temporarily. One of the most appealing aspects of this mechanism is its instant reaction time, which is not dependent on any specific technology. Figure 4.5 showcases an example of shifting workload in a data center for backing up purposes from 9:00 AM to 3:00 PM.

Migration

The corresponding mechanism involves the geographical shifting or moving of workload from one data center to another. For this mechanism to be effective, it is essential that the source and destination data centers are technically compatible. Specifically, both data centers should have the capability to execute the migrated workload and be equipped with virtualization technology. To this end, virtualization refers to the process of creating virtual instances or representations of physical resources, such as servers, storage devices, or networks.

This technology allows multiple virtual machines (VMs) or virtualized environments to run on a single physical server, effectively maximizing the utilization of hardware resources. It abstracts the underlying physical infrastructure, providing flexibility, scalability, and resource optimization. Users can allocate and scale virtual resources based on their specific needs, without being constrained by the limitations of physical hardware. Data centers commonly implement virtualization technologies using either open-source hypervisors like KVM (Habib 2008) and XEN (Barham et al. 2003), or licensed ones such as VMware (Rodríguez-Haro et al. 2012).

Figure 4.6 presents a comparison between two servers, one implementing virtualization (on the right) and the other without it (on the left). A physical server consists

Fig. 4.5 One example of workload shifting in a data center is when data backing up is rescheduled from the planned time of 9:00 AM, where there could be a potential peak-demand situation, to 3:00 PM during the off-peak period

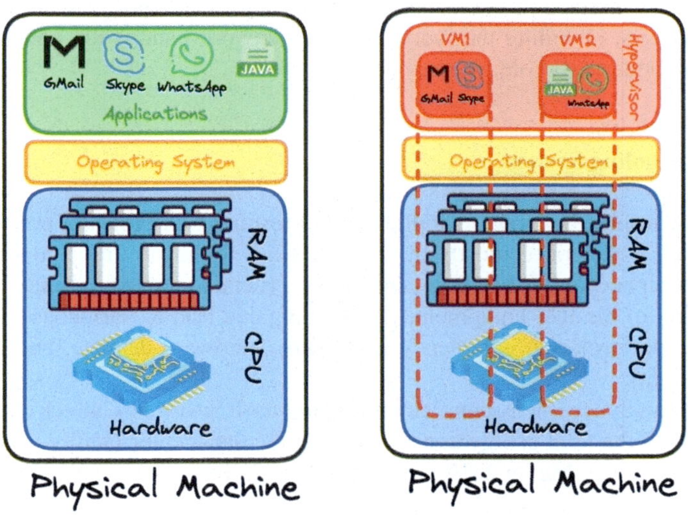

Fig. 4.6 The left figure demonstrates how applications like Skype, GMail, WhatsApp, and Java utilize hardware resources such as CPU and RAM through the operating system (OS). On the other hand, the right figure illustrates the concept of virtualization, where applications running in a specific virtual machine (VM) exclusively occupy hardware resources dedicated to that VM (denoted by red dotted lines). This is made possible through hypervisors running on top of the host machine's operating system. Each VM can have its own operating system, known as the guest OS, which manages the assigned resources like virtual CPU and virtual RAM. Note that the guest OS should not necessary be the same as the host OS

of hardware components like CPU for processing and RAM for data storage and retrieval. Applications such as GMail, Skype, WhatsApp, etc. utilize these hardware resources through the operating system, making it impossible to migrate these applications from one physical server to another. However, launching the same application from scratch on a different physical server is still feasible.

The physical server on the right side of the figure incorporates the virtualization concept using a hypervisor. The hypervisor operates as an application on top of the host server's operating system (OS) and enables the creation of virtual machines (VMs). In the given example, two VMs are created. Each VM has its own dedicated virtual resources, including CPU and RAM, which are part of the physical resources.

Similar to a physical server, each virtual machine also has its own OS, known as the guest OS, which can differ from the host OS. For instance, in the provided example, VM1 and VM2 can have Fedora and Ubuntu OS, respectively, while the hosting machine runs on Windows OS. The responsibility of the guest OS is to manage the applications running in that given VM. Thus, it is now possible to migrate applications from one physical server to another given the fact that those are running inside VMs.

This mechanism helps distribute the burden on power demand from one geographical site to the hosting data centers. However, it is important to note that this mechanism can have an impact on service downtime. In cases of live migration, the downtime can be minimized to seconds in optimal scenarios, while non-live migration can take minutes to hours. The reaction time of this mechanism depends on various factors, including the workload size, network bandwidth, and the chosen migration mode (e.g., cold, warm, live).

4.4.2.2 Cooling System

In ICT environments, power supplied to the resources is typically converted into heat, which is generated mostly by the CPUs due to performing computations. In densely equipped spaces like data centers, it is crucial to remove this heat to prevent overheating of the ICT equipment. Overheating can lead to various issues such as decreased performance, emergency shutdowns, hardware damage, and premature system failure.

To address this problem, data centers employ cooling systems, which can consume up to 50% of the total energy consumption of the data center facility in the case of HVAC technology (Meijer 2010). In Sect. 4.2, three distinct cooling technologies are discussed: air cooling, liquid cooling, and free cooling. The air cooling technique, also known as HVAC system, involves the use of CRAC and CRAH units, which are extremely inefficient from the energy point of view. This traditional method of cooling has been enhanced with a free cooling mode that leverages outside air.

Additionally, liquid cooling technology is often combined with HVAC systems to benefit from the superior specific heat capacity of liquid coolants, such as water, in comparison to air. These combined techniques (e.g., air with free cooling or air with liquid cooling) are referred to as hybrid cooling. The availability of various cooling technologies has led to numerous optimizations in the literature, focusing on utilizing the cooling system in an energy-efficient manner.

An optimization opportunity in this context involves adjusting the temperature setpoints within the recommended limits set by ASHRAE. This adjustment helps to reduce the power demand and energy consumption required for cooling, particularly in HVAC systems. ASHRAE regularly provides recommendations for temperature ranges, which currently range from 10 to 45 °C (Ni and Bai 2017), with a tendency to increase the upper limit. This difference of 35 °C between lower and upper bounds provides substantial flexibility and potential for energy savings in cooling operations.

One favorable scenario for implementing this mechanism in the context of HVAC systems is when there is an excess of energy from renewable sources. In such cases, the data center operator (DCO) can lower the operational temperature setpoint from 25 to 12 °C of the HVAC system. Hence, instead of curtailing the excess energy, it is used by the cooling system and is stored as a thermal buffer. During future demand-response periods, such as direct load control, the DCO can temporarily turn off the HVAC system until the critical setpoint of 45 °C is reached.

The time required to reach this critical temperature setpoint depends on factors like the size of the data center and its utilization rate. By adjusting the temperature setpoints, data centers can achieve a notable reduction in power demand. A recent example is Google, which increased the temperature setpoints in its data centers, resulting in decreased energy consumption (Google 2024). It has been suggested that DCOs can potentially save up to 5% on energy costs by raising the temperature setpoint by 1 °C (El-Sayed et al. 2012).

A new approach has emerged for implementing cooling mechanisms in data centers, which involves combining air cooling technologies with the utilization of cold regions like Finland. This strategy takes advantage of the naturally cold outside air to cool down the heat generated by ICT resources in data centers (Huang et al. 2020; Wahlroos et al. 2017). Furthermore, there is a suggestion to transfer the heat expelled from the data centers into district heating networks, providing free heating to residential areas.

However, it is important to note that this approach is still in its early stages and requires the establishment of the district heating network which is costly. Additionally, the process of heat transfer is not yet highly efficient, as there can be significant losses in the process. The reaction time of this mechanism is influenced by the size of the data center and the cooling technology employed.

Generally, an HVAC system can be deactivated immediately, but it may take approximately 10 min to cool down or heat up a specific data center. One key benefit of this mechanism is that it can be utilized without affecting the underlying SLAs. In fact, it can be integrated with other mechanisms mentioned earlier, particularly workload (refer to Sect. 4.4.2), to enhance flexibility and opportunities for demand-response and demand-side management.

4.4.2.3 Uninterruptible Power Supply

Traditionally, uninterruptible power supply (UPS) units have been incorporated into the ICT infrastructure of data centers to serve as a backup power source in the event of power outages or disconnection from the grid. As a result, most data centers are equipped with UPS units that can provide power to their ICT resources until either the local diesel generator takes over or alternative external power sources become available.

Despite their conventional use, UPS units can also be utilized to supply power to a data center for a limited duration, typically between 5 and 10 min, during periods of demand-response events and demand-side management. Furthermore, they can function as energy storage systems (ESS) during situations of excess power by charging their batteries.

To implement this mechanism effectively, the energy management system (EMS) of a data center must have control over the UPS system. The EMS sends control signals to the UPS to determine the most appropriate mode of operation. Additionally, models that account for battery capacity degradation are necessary to accurately assess the impact of charging and discharging the battery outside of its normal operating modes (Anseán et al. 2019; Severson et al. 2019; Wu et al. 2022; Yang et al. 2018).

As a result, this mechanism aids in reducing peak power demand and, in some cases, allows for temporary disconnection from the power grid. The reaction time of this mechanism is immediate, but its feasibility relies heavily on the battery's capacity, current state of charge (SoC), and state of health (SoH), as well as the utilization rate and size of the data center. Similar to the previous mechanism of the cooling system, the advantage of this mechanism is that it does not impact service level agreements (SLAs) when exploited by the EMS.

4.5 Summary

In the previous two chapters, the concept of Smart Grid and the principles of demand-side management (DSM) were introduced. This chapter takes a closer look at the potential of implementing DSM within the context of data centers. Data centers play a significant role as both major consumers of electricity and providers of flexibility. The chapter highlights key components of data centers, including ICT resources like servers, storage devices, and networking equipment, as well as cooling systems such as HVAC and backup power systems like UPS and diesel generators.

Then the concept of power usage effectiveness (PUE) is introduced as a state-of-the-art metric to evaluate the efficiency of data centers. The chapter demonstrates that energy management can be implemented in data centers, resulting in reduced energy consumption of ICT resources. This reduction in energy usage also indirectly decreases the need for cooling, making data centers more energy-efficient with lower

PUE values. The importance of energy management is illustrated through the example of Google data centers.

Following that, the concept of demand flexibility was explained and divided into two categories: time flexibility and amount flexibility. Subsequently, the mechanisms that offer flexibility in data centers were introduced, which can be classified as workload, cooling system, and UPS. It was emphasized that implementing the cooling system and UPS mechanisms does not affect the service level agreements (SLAs) of the active services. However, the workload flexibility mechanisms, including consolidation, migration, shifting, and DVFS, do impact the performance of the active service, consequently influencing the SLAs. Hence, it was highlighted that for such types of services and DSM, there is a need for relaxed SLAs.

References

Anseán D, García VM, González M, Blanco-Viejo C, Viera JC, Pulido YF, Sánchez L (2019) Lithium-ion battery degradation indicators via incremental capacity analysis. IEEE Trans Ind Appl 55(3):2992–3002

Barham P, Dragovic B, Fraser K, Hand S, Harris T, Ho A, Neugebauer R, Pratt I, Warfield A (2003) Xen and the art of virtualization. ACM SIGOPS Oper Syst Rev 37(5):164–177

Barroso LA, Hölzle U, Ranganathan P (2019) The datacenter as a computer: designing warehouse-scale machines, Chapter 6: Modeling Costs. Springer Nature

Basmadjian R, Botero JF, Giuliani G, Hesselbach X, Klingert S, De Meer H (2016) Making data centers fit for demand response: introducing greensda and greensla contracts. IEEE Trans Smart Grid 9(4):3453–3464

Basmadjian R, Niedermeier F, De Meer H (2012) Modelling and analysing the power consumption of idle servers. In: 2012 sustainable internet and ICT for sustainability (SustainIT), pp 1–9. IEEE

F. E. R. Commission et al. Assessment of demand response and advanced metering (2008)

El-Sayed N, Stefanovici IA, Amvrosiadis G, Hwang AA, Schroeder B (2012) Temperature management in data centers: why some (might) like it hot. In: Proceedings of the 12th ACM SIGMETRICS/PERFORMANCE joint international conference on measurement and modeling of computer systems, pp 163–174

Ghatikar G, Bienert R (2011) Smart grid standards and systems interoperability: a precedent with openadr. In: proceedings of the Grid interop forum, pp 1–7

Ghatikar G, Ganti V, Matson N, Piette MA (2012) Demand response opportunities and enabling technologies for data centers: findings from field studies. Technical report, Lawrence Berkeley National Lab.(LBNL), Berkeley, CA (United States)

Ghatikar G, Piette MA, Fujita S, McKane A, Dudley JH, Radspieler A, Mares K, Shroyer D (2009) Demand response and open automated demand response opportunities for data centers. Technical report, Lawrence Berkeley National Lab.(LBNL), Berkeley, CA (United States)

Google. Google data centers. https://www.google.com/about/datacenters/efficiency/#:~:text=How%20we%20do%20it,cooling%2C%20and%20build%20custom%20servers. Accessed 14 Feb 2024

Habib I (2008) Virtualization with kvm. Linux J 2008(166):8

Horner N, Azevedo I (2016) Power usage effectiveness in data centers: overloaded and underachieving. Electr J 29(4):61–69

Huang P, Copertaro B, Zhang X, Shen J, Löfgren I, Rönnelid M, Fahlen J, Andersson D, Svanfeldt M (2020) A review of data centers as prosumers in district energy systems: renewable energy integration and waste heat reuse for district heating. Appl Energy 258:114109

Katz RH (2009) Tech titans building boom. IEEE Spectrum 46(2):40–54
Koronen C, Åhman M, Nilsson LJ (2020) Data centres in future European energy systems-energy efficiency, integration and policy. Energ Effi 13(1):129–144
Li L, Zheng W, Wang X, Wang X (2014) Coordinating liquid and free air cooling with workload allocation for data center power minimization. In: 11th international conference on autonomic computing (ICAC 14), pp 249–259
Mämmelä O, Majanen M, Basmadjian R, De Meer H, Giesler A, Homberg W (2012) Energy-aware job scheduler for high-performance computing. Comput Sci-Res Dev 27:265–275
Meijer GI (2010) Cooling energy-hungry data centers. Science 328(5976):318–319
Ni J, Bai X (2017) A review of air conditioning energy performance in data centers. Renew Sust Energy Rev 67:625–640
Pore M, Abbasi Z, Gupta SK, Varsamopoulos G (2015) Techniques to achieve energy proportionality in data centers: a survey. Handbook on data centers, pp 109–162
Quan DM, Basmadjian R, De Meer H, Lent R, Mahmoodi T, Sannelli D, Mezza F, Telesca L, Dupont C (2012) Energy efficient resource allocation strategy for cloud data centres. In: Computer and information sciences II: 26th international symposium on computer and information sciences, pp 133–141. Springer
Rodríguez-Haro F, Freitag F, Navarro L, Hernánchez-sánchez E, Farías-Mendoza N, Guerrero-Ibáñez JA, González-Potes A (2012) A summary of virtualization techniques. Procedia Technol 3:267–272
Severson KA, Attia PM, Jin N, Perkins N, Jiang B, Yang Z, Chen MH, Aykol M, Herring PK, Fraggedakis D et al (2019) Data-driven prediction of battery cycle life before capacity degradation. Nat Energy 4(5):383–391
Shehabi A, Smith S, Sartor D, Brown R, Herrlin M, Koomey J, Masanet E, Horner N, Azevedo I, Lintner W (2016) United states data center energy usage report
Snowdon DC, Ruocco S, Heiser G (2005) Power management and dynamic voltage scaling: myths and facts. In: Proceedings of the 2005 workshop on power aware real-time Computing, New Jersey, USA, vol 31, p 34
Suleiman D, Ibrahim M, Hamarash I (2005) Dynamic voltage frequency scaling (dvfs) for microprocessors power and energy reduction. In: 4th international conference on electrical and electronics engineering, vol 12
Visual-Capitalist. Ranked: Top 50 data center markets by power consumption. https://www.visualcapitalist.com/cp/top-data-center-markets/. Accessed 17 Jan 2024
Wahlroos M, Pärssinen M, Manner J, Syri S (2017) Utilizing data center waste heat in district heating-impacts on energy efficiency and prospects for low-temperature district heating networks. Energy 140:1228–1238
Webb M et al (2008) Smart 2020: enabling the low carbon economy in the information age. The Climate Group. London 1(1):1–1
Wu Y, Liu Z, Liu J, Xiao H, Liu R, Zhang L (2022) Optimal battery capacity of grid-connected pv-battery systems considering battery degradation. Renewable Energy 181:10–23
Yang F, Wang D, Zhao Y, Tsui K-L, Bae SJ (2018) A study of the relationship between coulombic efficiency and capacity degradation of commercial lithium-ion batteries. Energy 145:486–495
Zhang Q, Meng Z, Hong X, Zhan Y, Liu J, Dong J, Bai T, Niu J, Deen MJ (2021) A survey on data center cooling systems: technology, power consumption modeling and control strategy optimization. J Syst Architect 119:102253

Chapter 5
Orchestration to Optimize Energy and Demand

5.1 The Process of Feedback Control

In Sect. 1.2, a cyber-physical system (CPS) is described as a network of interconnected subsystems utilizing information and communication technologies (ICT). The widespread integration of IoT devices, such as sensors, within CPS infrastructure, combined with advanced ICT capabilities, has facilitated access to vast amounts of data. This data abundance enables the establishment of a feedback control loop that plays a crucial role in shaping the future behavior of our digital environment.

Human-centered cyber-physical systems (HCPS) incorporate human operators (e.g., grid, data center, charging station) into the control loop, establishing a mutually influential relationship between humans and systems. Consequently, a specific process with well-defined steps is crucial for establishing a feedback control loop within an HCPS. This section provides a broad summary of this generic process and its steps (annotated from 1 to 5 in Fig. 5.1), emphasizing the significance of each in establishing this loop.

The process begins with the assessment of the environment by considering the foundational hardware infrastructure as well as the human operators and actors involved, as illustrated in Fig. 5.1. This hardware infrastructure integrates sensors and IoT devices as mentioned previously, aiming to continuously monitor the underlying infrastructure and generate raw data (annotated as 1). Raw data holds little value without extracting meaningful insights from it. Therefore, the subsequent step (annotated as 2) in this process involves leveraging the collected data.

Various methodologies have been proposed in the literature to preprocess raw data, including cloud, fog, or edge computing, where machine learning (ML) and Big Data techniques are utilized to create forecasts and prediction models. The primary distinction between cloud and fog/edge computing lies in how data transmission and computation are executed. In cloud computing, sensors and IoT devices transmit raw information at specified intervals to servers within cloud data centers.

R. Basmadjian, *Demand-Side Energy and Power Management in Data Centers*, Lecture Notes in Energy 1, https://doi.org/10.1007/978-3-031-85956-4_5

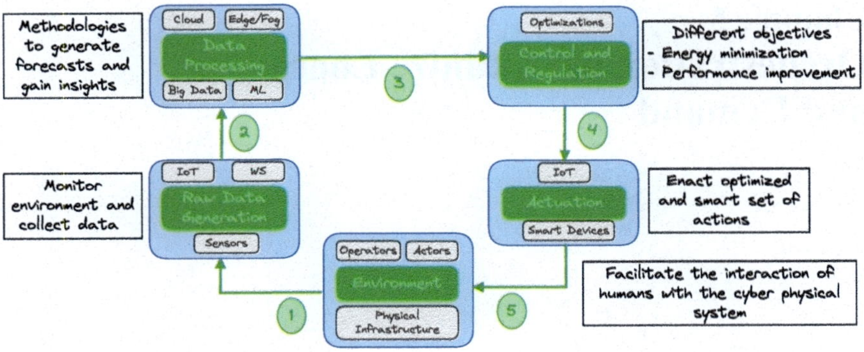

Fig. 5.1 The feedback control loop begins with monitoring the environment, including physical infrastructure and human operators, to gather raw data. This data is processed using cloud, edge, or fog computing, along with Big Data and machine learning, to generate valuable insights and forecasts. These insights feed into optimization engines aimed at improving control of systems like cyber-physical environments, often balancing conflicting objectives such as demand reduction and performance enhancement. Once the optimal configuration is determined, it is implemented via actuating devices like IoT or smart equipment, emphasizing the importance of effective human-machine interfaces (HMI) for user interaction with the system

This necessitates high communication bandwidth for IoT devices and powerful servers to process the vast data volumes from numerous sensors and IoT devices in the environment. Conversely, in fog and edge computing, IoT devices and sensors either possess computing capabilities themselves or are in close proximity to powerful equipment for data processing, reducing transmission bandwidth usage and alleviating stress on cloud data center servers.

The forecasts and predictions generated will be incorporated with real-time data coming from the IoT devices deployed in the infrastructure to effectively control and manage the underlying environment (annotated as 3). To accomplish this, control algorithms are bolstered by optimization engines that consider multiple objectives, some of which may conflict. For instance, balancing the reduction of server power demand with enhancing service performance presents a potential conflict. The primary goal of these control algorithms, supported by optimization engines, is to determine the most optimal configuration.

This optimal setup must adhere to two key properties which themselves are competing: responsiveness and stability. These properties, rooted in control theory, dictate that actions taken by those algorithms should prevent uncontrollable oscillations (stability) and enable the infrastructure to maintain performance goals despite changes in its operational environment (responsiveness). Control theory provides powerful tools for building responsive and stable systems whose details are out of the scope of this chapter and book. Interested readers can find a detailed explanation of control theory in Chap. 8 of Keshav (2012).

The optimal actions determined by the optimization engine must be implemented back into the environment using actuating devices (annotated as 4) like IoT and smart equipment (see Sect. 3.4.4). Examples of these actions include decisions such as "charging the EV with the lowest power due to detected fluctuations in the grid's voltage and frequency levels" or "migrating a VM from an energy-intensive server to an energy-efficient one and shutting down the initial server". It is crucial for these actions to be transparent to human actors and operators, given the interdependence between humans and complex systems like HCPS as mentioned previously (annotated as 5).

This transparency is achieved through well-designed human-machine interfaces (HMI), although specific details are beyond the scope of this chapter and book. In Myers (1998), readers can explore a concise history of HMI, while Hartson and Hix (1989) offers a thorough survey on various aspects like dialogue independence, development methods, and control structures, among others. Additionally, Guccione and McKirahan (2016) introduces key concepts in HMI and includes practical lab exercises to enhance the understanding of creating and operating process control applications.

To summarize, in this section, we have observed how the feedback control loop efficiently concludes by first monitoring the environment's hardware infrastructure through monitoring tools using IoT devices and concluding with the execution of optimal actions back into the environment by actuating devices such as IoT or smart equipment, while relaying feedback to human operators through well-designed HMI.

5.2 Energy Management System

5.2.1 Essential Components

An energy management system (EMS) is a software-based orchestrator that helps entities and organizations monitor, control, and optimize their electricity (e.g., demand and energy) usage. As such, the corresponding EMS needs to be designed considering the process of the feedback control loop of Sect. 5.1. Thus, it collects data from the underlying infrastructure and analyzes this data to identify opportunities for energy savings (see Sect. 5.3) as well as demand adaptation (see Sect. 5.4), for energy efficiency improvements and demand-response advancement respectively.

By providing real-time insights into energy consumption as well as power demand patterns, EMS enables organizations to make informed and controlled decisions about how to reduce energy and demand costs, minimize waste, and lower their carbon footprint. Additionally, EMS can help automate energy- and demand-intensive processes, schedule equipment usage more efficiently, and track energy and demand performance over time.

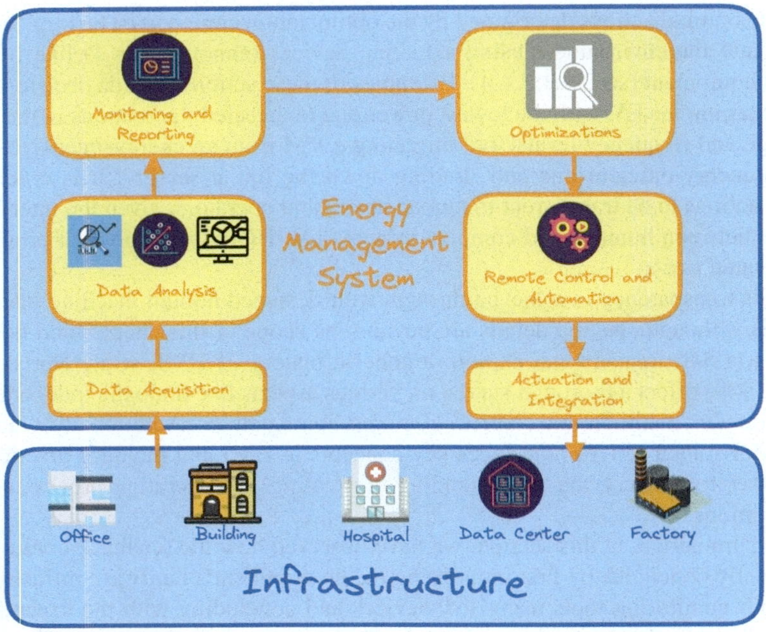

Fig. 5.2 An energy management system (EMS), which is a software-based solution, encompasses essential functions including data collection, analysis, reporting, optimization, remote control, automation, actuation, and integration. The corresponding EMS can be integrated with different environments and infrastructures such as offices, buildings, hospitals, data centers, and factories, among others

Figure 5.2 showcases the most relevant tasks that a software-based orchestrator such as an EMS needs to realize and implement. It typically consists of several key components, including:

- Data acquisition: EMS collects data from various sources, such as smart meters, sensors, and building management systems, to monitor energy consumption and power demand in real-time.
- Data analysis: The collected data is analyzed to identify patterns, trends, and anomalies in both demand and energy usage. This analysis helps organizations understand where energy (power) is being consumed (demanded) and where potential savings (adaptations) can be made.
- Monitoring and reporting: EMS provides real-time monitoring of energy consumption and power demand, as well as generates reports and dashboards to visualize both energy and power usage data. This allows organizations to track their energy and demand performance and make data-driven decisions.
- Energy and demand optimization: EMS uses advanced algorithms and predictive analytics to optimize energy as well as demand usage and reduce costs.

This may involve adjusting equipment schedules, setting energy-saving targets, or implementing demand-side management strategies.
- Remote control and automation: Some EMSs offer remote control capabilities, allowing users to adjust equipment settings and control energy and demand usage from a centralized platform. Automation features can help streamline energy- and power-intensive processes and improve overall efficiency.
- Actuation and integration with other systems: EMS needs to switch to the optimal configuration through actuation. It can be also integrated with other building management systems, such as HVAC, lighting, and security systems, to create a more holistic approach to energy management. This integration enables organizations to optimize energy and demand usage across multiple systems and devices.

Overall, an EMS serves as a comprehensive solution for organizations to monitor, analyze, and optimize their energy and demand usage, leading to cost savings, enhanced sustainability, and compliance with regulatory standards. It is important to note that when designing an EMS for Smart Grid applications, it is crucial to meet interoperability requirements by adhering to the guidelines outlined in SGAM (see Sect. 2.2.2). Furthermore, an EMS can be seamlessly integrated into various settings such as hospitals, buildings, offices, factories, and data centers, among others. In the following sections, examples of EMS implementations for managing energy consumption and power demand in data centers will be discussed.

5.2.2 Multi-agent Implementation

To implement a software-based orchestrator and bring the concept of EMS to life, the various subsystems within the underlying HCPS need to be interconnected and capable of working together to achieve the goals outlined by the EMS. Research has highlighted the potential and benefits of utilizing agents to achieve this level of coordination and collaboration.

An agent is a software program or entity that operates autonomously to carry out tasks or make decisions on behalf of a user or another system. Agents are designed to function independently, often possessing intelligence or learning capabilities to accomplish specific objectives. They can engage with their surroundings, other agents, or users through communication channels and sensors, enabling them to collect information, analyze data, and take actions based on predefined rules or algorithms.

The multi-agent paradigm goes beyond just separating concerns or abstracting software. It offers a valuable methodological approach for modeling intricate real-world scenarios that arise from the interactions of multiple decision-making entities. As these entities may have differing values and objectives, the multi-agent paradigm enables the modeling of systems with competing agents. Additionally, it equips us with tools to comprehend the behavior of these entities and provides mechanisms to regulate their interactions, ensuring the fulfillment of their respective values and goals. A detailed introduction to multi-agent systems can be found in Wooldridge (2009).

Given the effectiveness and adaptability of multi-agent systems in addressing complex problems such as optimal resource allocation, the fundamental components that define these systems are allocation protocols and strategies. Allocation protocols facilitate information exchange among agents, while allocation strategies consist of rules that guide individual agents in determining their specific responses within the allocation protocols.

These two components closely collaborate: agents utilize allocation strategies to pursue the best allocation outcome from their own standpoint, while the allocation protocol governs the structure and content of interactions among agents based on environment-defined rules to achieve a desired collective result. By separating the allocation protocol from the allocation strategies, multi-agent resource allocation mechanisms gain the capability and flexibility to adjust to diverse circumstances effectively.

The contract net protocol (CNP) (Smith 1980) is a state-of-the-art allocation protocol used predominantly in multi-agent software systems like EMS to facilitate task allocation and coordination among autonomous agents. It involves a bidding process where agents can announce their capabilities and bid on tasks that need to be completed. The protocol helps in efficiently assigning tasks to agents based on their abilities and availability, ultimately improving the overall performance of the system. It functions as a specific type of auction mechanism, where one party (known as the initiator) solicits a particular service from multiple suppliers (referred to as participants).

The agents involved in the protocol are driven by self-interest as they strive to achieve their individual objectives. Should an agent cease operation, such as due to a communication failure, the system as a whole will persist, allowing the respective agent to rejoin later. Various system models can be applied within CNP. For example, it can be assumed that communication is reliable, with messages potentially experiencing delays but ultimately being delivered in the order they were sent. Processing time can be controlled through timeouts in the protocol, while the volume of communication can be managed by the throughput of the communication channels in use.

Figure 5.3 depicts the interactions between the initiator and participants in CNP. The initiator shares details of the requested service through a call for proposals (e.g., task announcement). Participants can either decline (if unable to provide the service) or submit a service proposal (e.g., task bidding) along with a corresponding compensation (e.g., in the form of a cost) for executing the requested service for the initiator.

Following the receipt of refusals or proposals from all participants within a specified deadline, the initiator evaluates the proposals to determine the most suitable ones based on how well the compensation aligns with the initiator's values and objectives (e.g., selecting the lowest cost) and communicates the chosen proposals

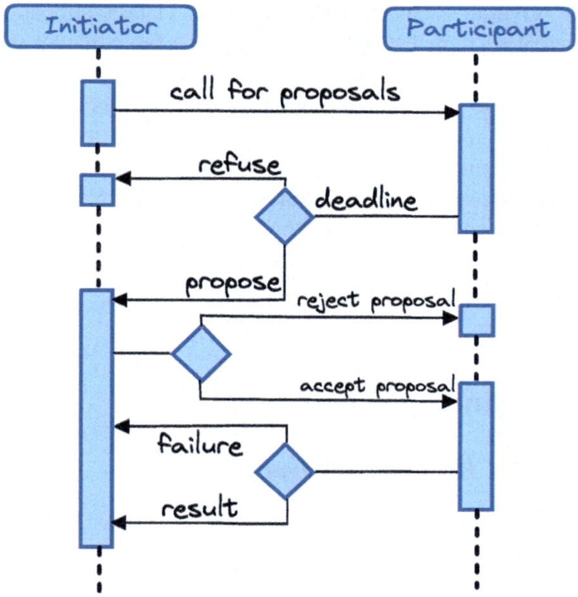

Fig. 5.3 The flowchart diagram between the initiator and participant describing the interactions of those entities within the contract net protocol (CNP) (Smith 1980)

to the participants (e.g., task assignment). Subsequently, participants whose proposals are accepted inform the initiator about completing the service(s) or any issues encountered (e.g., task execution).

Thus, the CNP helps in efficient task allocation and coordination in multi-agent systems by allowing agents to negotiate and collaborate on tasks autonomously. It offers several advantages in multi-agent systems. Some of these include:

- Efficient task allocation: The protocol allows for a decentralized and dynamic allocation of tasks based on agent capabilities and availability, leading to more efficient task completion.
- Flexibility: Agents have the flexibility to bid on tasks based on their strengths and preferences, leading to a more tailored and optimized task assignment process.
- Scalability: The protocol can easily scale to accommodate a large number of agents and tasks, making it suitable for complex multi-agent systems.

However, there are also some inconveniences associated with the CNP:

- Communication overhead: The bidding and negotiation process can introduce communication overhead, especially in systems with a large number of agents, potentially impacting system performance.
- Lack of global optimization: The protocol focuses on local task allocation decisions based on individual bids, which may not always lead to globally optimal task assignments.
- Vulnerability to strategic behavior: Agents may strategically manipulate their bids to gain a competitive advantage, potentially leading to sub-optimal task allocations.

Overall, while the CNP benefits in task allocation and coordination in multi-agent systems like EMS, it is important to consider these potential drawbacks when implementing the protocol. One example of a project that has used the CNP is the RoboCup Rescue Simulation League. In this project, teams of autonomous agents (robots) are tasked with simulating disaster response scenarios, such as search and rescue missions in a simulated disaster area. The CNP allocates tasks to different robots based on their capabilities and availability, allowing for efficient coordination and collaboration among the robot teams. By using the protocol, the project aims to improve the overall performance and effectiveness of the robot teams in responding to disaster scenarios.

5.3 Energy Minimization

5.3.1 High-Level Overview

The primary goal of implementing this type of energy management system (EMS) is to reduce the energy consumption of the data center while maintaining the service level agreements (SLAs) with its IT customers. This energy optimization strategy is focused on enhancing energy efficiency within the data center operations. The data center operator (DCO) is motivated to pursue this objective as energy consumption has emerged as a key factor impacting operational expenditure (e.g., Opex) (Barroso and Clidaras 2022; Koomey et al. 2007). By delivering the same quality of service (QoS) to customers while lowering energy usage, the data center can effectively decrease its monthly electricity expenses. As a result, the DCO can offer more cost-effective services, fostering increased competitiveness within the industry.

Figure 5.4 gives a high-level overview of the key components of such an EMS. Specifically, it comprises two main elements: “Demand Predictor” and “Optimizer”. Furthermore, the “Agnostic Orchestrator” plays the role of interfacing between those two elements and the underlying “Infrastructure”. The primary goal of the “Demand Predictor” (see Sect. 5.3.2) is to forecast the demand for ICT resources like servers, storage devices, and networking equipment based on scenarios provided by the “Optimizer” (see Sect. 5.3.3).

The “Optimizer” interacts with the “Demand Predictor” by posing questions such as “What will be the energy consumption of Server X with an additional workload of 30%?” or “What is the energy consumption when migrating VM 1 from Server 1 to Server 2?” These inquiries help the “Optimizer” determine the most optimal hardware configuration to minimize energy usage.

The “Demand Predictor” and “Optimizer” engage in a two-way exchange of information, allowing them to collaborate effectively. It is crucial for the “Optimizer” to have a clear understanding of the service level agreements (SLAs) associated with active services to optimize energy usage while meeting SLA-related key performance indicators (KPIs) and preventing any breaches.

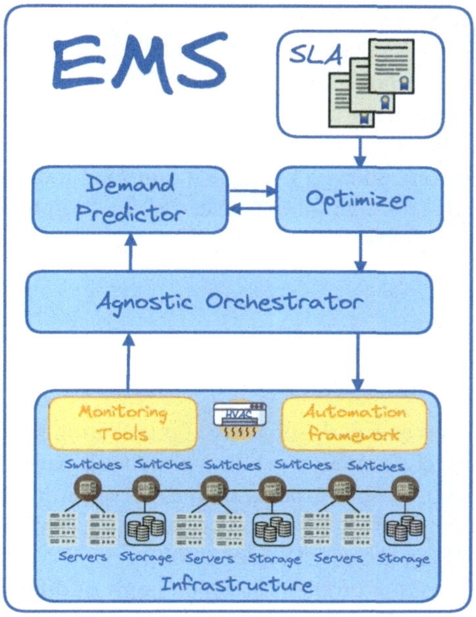

Fig. 5.4 A high-level overview of the key architectural components of an energy management system (EMS) designed to minimize the energy usage of a data center while upholding the service level agreements (SLAs) with IT customers. Two pivotal elements within this EMS are the "Demand Predictor" and the "Optimizer". The "Agonstic Orchestrator" plays a crucial role in establishing the interfaces between the data center's "Infrastructure" and the inner parts of the EMS such as the "Demand Predictor" and "Optimizer". The directional arrows indicate the flow of information between these components

5.3.2 Demand Predictor

It employs models to precisely forecast the power requirements of servers, storage devices, and networking equipment. These models are crucial in accurately depicting the components that demand the most power during periods of inactivity and activity. During idle times, the ICT resources are inactive, while during active periods, they are engaged in tasks like computation, communication, or storage operations. For example, a server's idle and dynamic power demands can be analyzed based on the CPU, memory, hard disk, and networking interfaces, among others. More detailed information on these demand estimation models will be provided in Chap. 6.

These models require both static and dynamic input data for accurate predictions. Static information remain constant throughout the optimization period, such as the number of CPU cores as well as memory modules, and memory capacity for a given server. On the other hand, dynamic information change frequently and need to be monitored, like CPU utilization rate, available memory capacity, and transfer rate of the networking interface, to name a few.

Fortunately, data centers do not need additional IoT devices for monitoring as ICT resources already have integrated monitoring and actuating capabilities accessible through tools like Nagios (Barth 2008; Josephsen 2007), Zabbix (Dalle Vacche 2015; Olups 2010), and Prometheus (Turnbull 2018), among others. These tools track the performance and health of servers, networks, and storage devices in real-time. The dynamic information collected by these monitoring tools is then transmitted to the "Demand Predictor" through the interfaces of the "Agnostic Orchestrator" (see Sect. 5.3.4) for accurate predictions.

5.3.3 *Optimizer*

It is typically formulated as a multi-objective optimization problem, where the goal is to minimize energy consumption while maintaining performance (e.g., SLAs). This involves optimizing various aspects such as server utilization, cooling systems efficiency, workload scheduling, and resource allocation. This optimization problem is similar to the one of bin packing (Seiden 2002).

It is a classic optimization problem in which items of different sizes (e.g., workload) must be packed into a limited number of bins or containers (e.g., ICT resources) in the most efficient way possible. The goal is to minimize the number of bins used while ensuring that the items fit within the capacity of each bin. Figure 5.5 gives a graphical illustration of the corresponding items, a container, and the packing of those items in the given container.

The bin packing problem is known to be NP-hard (Fekete and Schepers 2001), which means that there is no known polynomial-time algorithm that can solve it optimally for all instances. However, there are approximation algorithms (Coffman

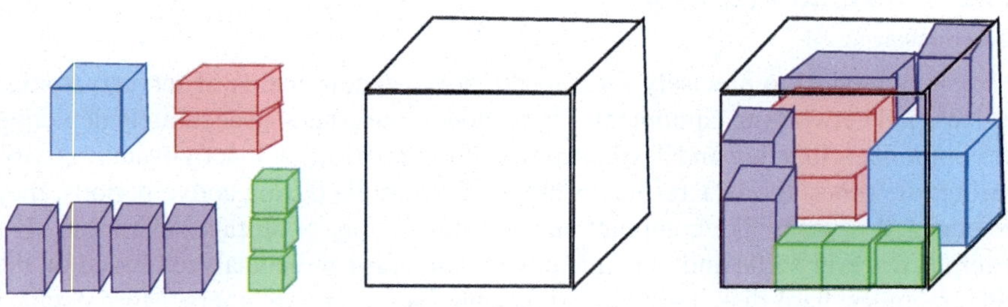

(a) Four item types. **(b)** A container. **(c)** Packing of items to the container.

Fig. 5.5 **a** In this 3D bin packing example, there are four different types of items: blue (1), red (2), green (3), and purple (4), totaling 10 items. Each type of item corresponds to specific resource requirements for workloads, such as CPU, memory capacity, and network bandwidth for instance. **b** An empty container symbolizing the available resources of a server. **c** The arrangement of these 10 items into the container. Please note that this presentation is for illustrative purposes and may not depict the most efficient packing solution

et al. 1984; Garey and Johnson 1981) and heuristics that can provide near-optimal solutions in a reasonable amount of time. Additionally, other optimization techniques such as mathematical programming as well as machine learning algorithms are used to find the best trade-offs between energy efficiency and performance.

The optimal configuration the “Optimizer” decides is applied to the “Infrastructure” using the automation frameworks already deployed in data centers. Examples of such automation frameworks include Ansible (Hochstein and Moser 2017), Puppet (Benson et al. 2019; Frank et al. 2017), Chef (Taylor and Vargo 2014), and SaltStack (Hall 2016), to name a few. Ansible is an open-source tool for configuration management, application deployment, and task automation. It uses simple YAML-based playbooks to define automation tasks.

Puppet is another configuration management tool that automates the provisioning, configuration, and management of servers. It uses declarative language to define the desired state of systems. Chef is a powerful automation platform that uses code to automate infrastructure tasks. It allows for defining infrastructure as code and automating the deployment and management of servers. SaltStack is a scalable automation tool that uses a master-minion architecture to manage and automate infrastructure. It allows for remote execution, configuration management, and orchestration of systems.

5.3.4 Agnostic Orchestrator

As previously mentioned, the purpose of such a component is to act as the intermediary between the underlying hardware infrastructure and the internal functions of the EMS, like the “Demand Predictor” and “Optimizer”. The key characteristic of this orchestrator is its ability to remain independent of specific hardware technologies, data center types (e.g., cloud computing, high-performance computing), and software implementations (e.g., Nagios, Chef, Puppet, etc.). This agnosticism is achieved through the use of connectors. The corresponding modular concept of connectors simplifies the interaction between hardware (e.g., infrastructure) and software (e.g., EMS) by developing the necessary connectors.

Software-based solutions known as connectors are designed to create connections through application programming interfaces (APIs). In Fig. 5.6, the “Agnostic Orchestrator” features two types of connectors: North connectors linking it to the “Demand Predictor” and “Optimizer” within the EMS system, requiring only one connector for each component due to their internal implementation. On the other hand, south connectors establish connections between the “Agnostic Orchestrator” and various monitoring tools and automation frameworks.

Since there is no single standardized solution for monitoring and automation, specific connectors must be developed for tools like Nagios, Zabbix, Prometheus, as well as automation frameworks like Ansible, Puppet, Chef, and SaltStack, among others. These connectors enable the exchange of information and interaction between the

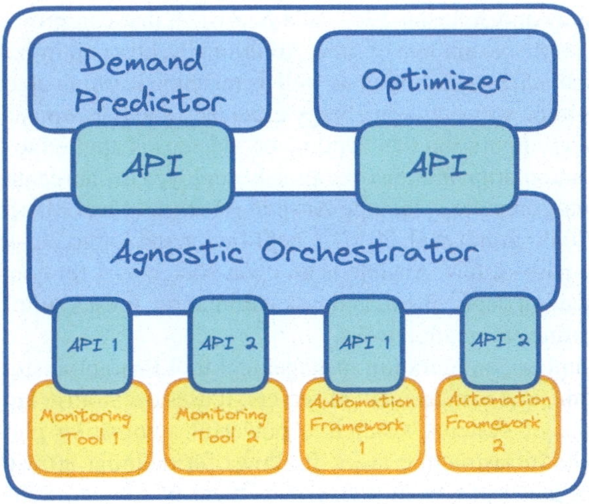

Fig. 5.6 The "Agnostic Orchestrator" provides two sets of connectors, known as north and south connectors (e.g., API). These connectors enable seamless communication and data exchange between the underlying monitoring tools and automation frameworks with the "Demand Predictor" and "Optimizer". It is important to note that north connectors only need to be set up once, as the "Demand Predictor" and "Optimizer" implementations are tailored and internal to the EMS. Conversely, south connectors require a unique setup for each monitoring tool and automation framework due to the existence of several independent tools and frameworks

orchestrator, monitoring tools, and automation frameworks, making the orchestrator agnostic.

5.3.5 Infrastructure

To optimize energy consumption by the corresponding EMS, it is crucial to have a detailed understanding of the data center's hardware and software infrastructure. The first step involves identifying all ICT resources within the data center together with the corresponding power supply as well as the cooling system, as shown in Fig. 5.7. This includes enclosures housing computing resources (such as physical servers, virtual machines, and containers), storage arrays (like NAS, SAN, and tape library systems), and networking devices (such as routers, switches, firewalls, and load balancers). Enclosures, depicted in Fig. 4.2, are cabinet-like structures made of metal that house ICT resources like servers, networking equipment, and storage devices.

Just like how a data center is illustrated in Fig. 5.7 with its hardware and software components, the same level of detail should be provided for each ICT resource. Figure 5.8 showcases an example focusing on servers within data centers. There are

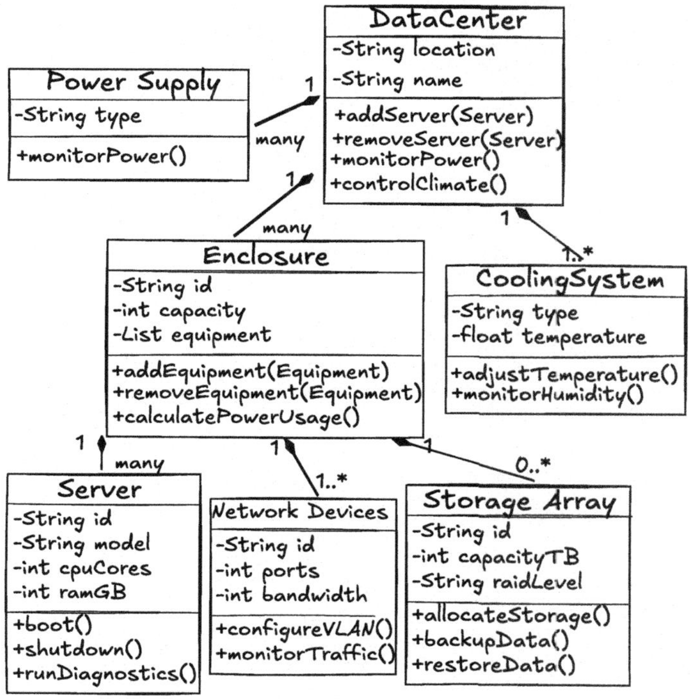

Fig. 5.7 A visual representation showcasing the three main components of enclosures, power supply and cooling system housed and overseen within a data center

three main types of servers: rack, blade, and tower. Rack servers are housed in a standard enclosure, while blade servers are compact and fit into a blade enclosure, enabling efficient management of multiple servers in a smaller area.

Blade servers typically share power, cooling, networking, and other resources within the enclosure, whereas rack servers function independently. Although blade servers are often more energy-efficient than rack servers, they may have limitations in terms of customization and expansion options. Blade servers can be configured to be hot-swappable, allowing for easier replacement in case of issues, and promoting redundancy.

A tower-like server is a type of server that is designed in a vertical tower form factor, resembling a traditional desktop computer tower. These servers are standalone units that are typically used in small office or home office environments where space is not a constraint. Tower servers are easy to set up and maintain, making them a popular choice for businesses with limited IT resources.

Every server, regardless of its specific type, comprises essential hardware and software components. From a hardware perspective, a server typically includes a motherboard that facilitates connectivity between different hardware components and provides circuitry, along with a protective metal chassis for these components. CPUs

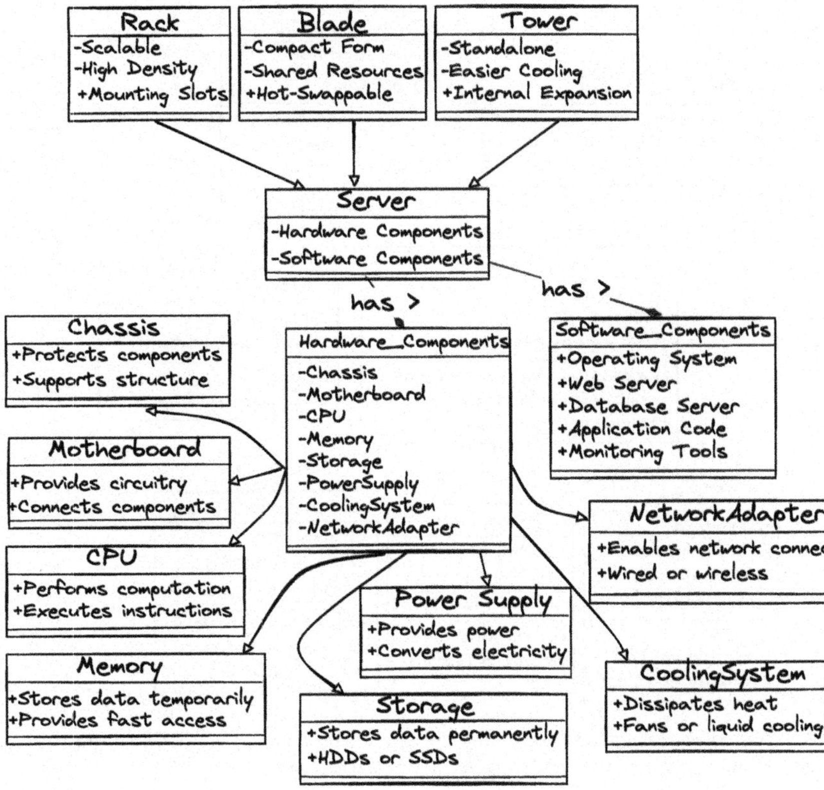

Fig. 5.8 Introduction to three server types: rack, blade, and tower servers. Each of these server types can be further broken down into their respective hardware and software components. The "+" symbol indicates that the field can be broken down into more detailed and specific components

are connected to the motherboard to execute instructions and perform computations, while memory modules like RAM store information temporarily for quick access. Storage devices such as hard disk drives (HDD) or solid-state disks (SSD) store information permanently.

Network adapters are also connected to the motherboard to enable network connectivity in both wired and wireless forms. Power supply units (PSUs) deliver electricity to electronic components, which generate heat during operations and require cooling to prevent premature failure. Cooling systems, utilizing liquid cooling technology or fans, are employed to dissipate this heat. On the software side, each server is equipped with an operating system like Windows, Linux, or Ubuntu, and runs web and/or database servers, monitoring tools, and various other applications.

To ensure clarity within the given space constraints, we will provide an additional example of breaking down the CPU into its various components, as shown in Fig. 5.9. In the case of a server's CPU, it may consist of one or more cores, each with shared

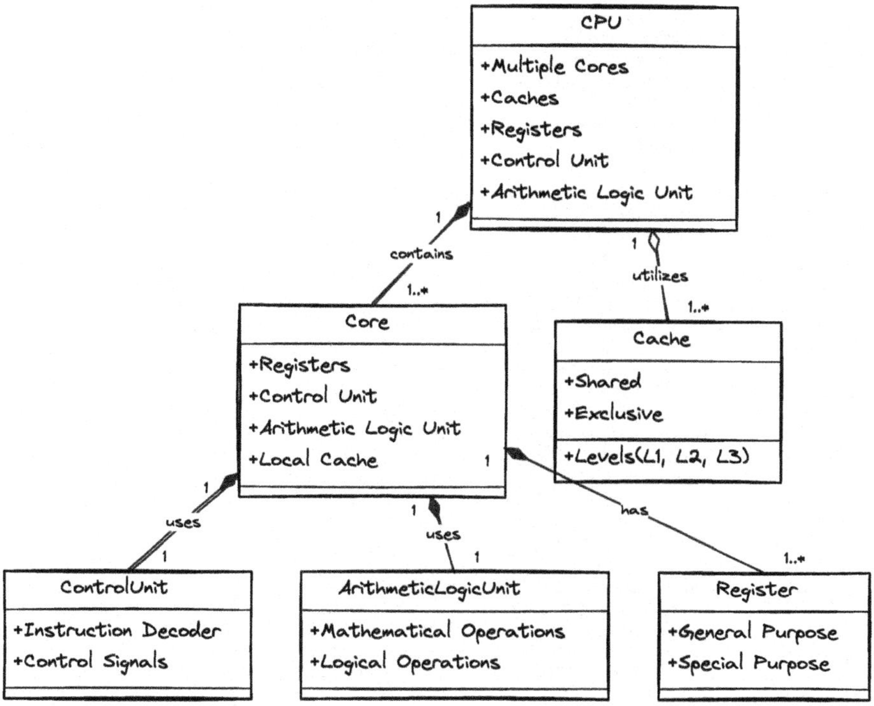

Fig. 5.9 Breaking down a multi-core central processing unit (CPU) into its fundamental components, including cores, caches, registers, arithmetic logic units, and control units. The "+" symbol indicates that the field can be broken down into more detailed and specific components

and exclusive caches at different levels (e.g., L1, L2, and L3). Furthermore, each core contains registers that can be general or special purpose, along with its local cache, control unit (CU), and arithmetic logic unit (ALU).

The control unit manages signals and instruction decoding, while the arithmetic logic unit is essential for executing mathematical (e.g., addition, subtraction, division, and multiplication) and logical (e.g., AND, OR, XOR, NOT, etc.) operations. It is important to note that this breakdown should be applied to all the ICT resources of a given data center illustrated in Fig. 5.7. Interested readers can find a detailed overview of a data center, its ICT resources, and the key demand-related attributes in Basmadjian et al. (2012).

After creating a comprehensive description of the data center and its components, the subsequent task involves pinpointing both static and dynamic parameters of each component essential for the power estimation models of the "Demand Predictor". It is crucial to identify these dynamic parameters to enable monitoring tools to furnish real-time data. Examples of dynamic parameters that server monitoring tools can offer encompass CPU utilization rate, the frequency and voltage of CPU cores, the available memory capacity, and network adapter transfer rates, among others.

```
<?xml version="1.0" encoding="UTF-8" standalone="no"?>
<xsd:schema xmlns:dcdr="http://www.dcdr.com/schemes/MetaModel"
  xmlns:xsd="http://www.w3.org/2001/XMLSchema"
  xmlns:jaxb="http://java.sun.com/xml/ns/jaxb"
  jaxb:version="2.0"
  targetNamespace="http://www.dcdr.com/schemes/MetaModel">
  <xsd:annotation>
    <xsd:appinfo>
      <jaxb:globalBindings mapSimpleTypeDef="true" />
    </xsd:appinfo>
  </xsd:annotation>

  <xsd:element  name="EMS" type="dcdr:EMSType"/>
  <xsd:element  name="Site" type="dcdr:SiteType"/>
  <xsd:element  name="Datacenter" type="dcdr:DatacenterType"/>
  <xsd:element  name="Rack" type="dcdr:RackType"/>
  <xsd:element  name="Server" type="dcdr:ServerType"/>
  <xsd:element  name="CPU" type="dcdr:CPUType"/>
  <xsd:element  name="RAMStick" type="dcdr:RAMStickType"/>
  <xsd:element  name="HardDisk" type="dcdr:HardDiskType"/>
  <xsd:element  name="StorageUnit" type="dcdr:StorageUnitType"/>
  <xsd:element  name="Mainboard" type="dcdr:MainboardType"/>
  <xsd:element  name="PSU" type="dcdr:PSUType"/>
  <xsd:element  name="Fan" type="dcdr:FanType" />
  <xsd:element  name="RAID" type="dcdr:RAIDType"/>
  <xsd:element  name="Cache" type="dcdr:CacheType"/>
  <xsd:element  name="Core" type="dcdr:CoreType"/>
  <xsd:element  name="SolidStateDisk" type="dcdr:SolidStateDiskType"/>
  <xsd:element  name="TowerServer" type="dcdr:TowerServerType"/>
  <xsd:element  name="BladeServer" type="dcdr:BladeServerType"/>
  <xsd:element  name="RackableServer" type="dcdr:RackableServerType"/>
```

Fig. 5.10 A meta-model template in XSD format that can be used to generate different data center instances. This part shows the different declared types of entities of the meta-model. As this example is for demonstration purposes, it may not show all the details

The one simple way to encode all this information is the usage of the extensible markup language (XML), where there are tools that convert the XML presentation to any other format such as Java script object notation (JSON). Figure 5.10 provides a template in XML schema definition (XSD) format that can be used to instantiate different presentations of data centers. The web address given on the second line from the left column (e.g., http://www.dcdr.com/schemes/MetaModel) is a fictitious one shown for demonstration purposes.

Figure 5.11 presents a partial continuation of the relevant XSD template. This excerpt provides further details and extensions to the schema, illustrating how elements, attributes, and data types are defined and organized within the XML structure. The root element is the EMS that can manage one or more sites, where each site's main power-related attributes are the carbon (CUE) and power usage effectiveness (PUE) as well as the power demand monitored (e.g., measured) and reported (e.g., computed power) by the installed IoT devices (e.g., Step 1 in Fig. 5.1). Also, notice that each site may include one or multiple data centers.

Figure 5.12 gives a detailed overview of the data center entity. To this end, each has its unique computing style (e.g., cloud, high-performance, traditional), power

```
  </xsd:restriction>
</xsd:simpleType>
<xsd:simpleType name="BitErrorRateType">
  <xsd:restriction base="xsd:double">
    <xsd:minInclusive value="0"/>
  </xsd:restriction>
</xsd:simpleType>
<xsd:simpleType name="ScenarioType">
  <xsd:restriction base="xsd:string">
    <xsd:enumeration value="single"/>
    <xsd:enumeration value="federated"/>
  </xsd:restriction>
</xsd:simpleType>
<xsd:simpleType name="LocationType">
  <xsd:restriction base="xsd:string" />
</xsd:simpleType>

<xsd:complexType name="EMSType">
  <xsd:sequence >
      <xsd:element ref="dcdr:Site" minOccurs="0" maxOccurs="unbounded"/>
      <xsd:element name="description" type="xsd:string" minOccurs="0" maxOccurs="unbounded"/>
      <xsd:element name="version" type="xsd:string" minOccurs="0" maxOccurs="unbounded"/>
  </xsd:sequence>
</xsd:complexType>
<xsd:complexType name="SiteType">
      <xsd:sequence >
      <xsd:element  name="PUE" type="dcdr:PUEType" minOccurs="1" maxOccurs="1"/>
        <xsd:element  name="CUE" type="dcdr:CUEType" minOccurs="1" maxOccurs="1"/>
      <xsd:element  name="computedPower" type="dcdr:PowerType" minOccurs="0" maxOccurs="1"/>
      <xsd:element ref="dcdr:Datacenter" minOccurs="0" maxOccurs="unbounded"/>
  </xsd:sequence>
</xsd:complexType>
```

Fig. 5.11 This part of the XSD template shows the different simple and complex types of entities of the meta-model. As this example is for demonstration purposes, it may not show all the details

demand (e.g., computed power), and a range of ICT resources like servers, networking equipment, and storage devices, along with framework capabilities. Note the distinction between different server types, including rack, blade, and tower servers, all falling under the main server category. Server attributes encompass details like name, status (e.g., idle, working), reported and estimated demand, motherboard connecting resources like CPU, memory, network cards, and hard disk, as well as native OS and hypervisor (not shown in Fig. 5.12).

Figure 5.13 presents an illustrative example of an instance created by considering the EMS template shown in Figs. 5.10, 5.11 and 5.12. This example features a single site with PUE and CUE values of 1.2 and an overall power demand of 288.2 kW. The site houses a cloud computing data center comprising two racks and leveraging Nagios as its framework capability. Further insights are provided in Fig. 5.14, offering a detailed breakdown of one of the racks.

This rack includes an enclosure housing a blade server, three PSUs, and six fans. The blade server is equipped with specific details such as its name, status, framework ID essential for EMS optimizations, computed and measured powers of 168.1 kW and 119 kW respectively, motherboard containing IT resources (not shown in the figure), a native Linux operating system, and a XEN host hypervisor.

This digital representation of a data center's hardware infrastructure, along with its software elements like monitoring tools, automation frameworks, operating systems,

```
<xsd:complexType name="DatacenterType" >
      <xsd:sequence >
      <xsd:element  name="computingStyle" type="dcdr:DCComputingStyleType" minOccurs="1"
maxOccurs="1"/>
      <xsd:element  name="computedPower" type="dcdr:PowerType" minOccurs="0" maxOccurs="1"/>
      <xsd:element ref="dcdr:Rack" minOccurs="0" maxOccurs="unbounded"/>
      <xsd:element ref="dcdr:TowerServer" minOccurs="0" maxOccurs="unbounded"/>
      <xsd:element ref="dcdr:BoxRouter" minOccurs="0" maxOccurs="unbounded"/>
      <xsd:element ref="dcdr:BoxSwitch" minOccurs="0" maxOccurs="unbounded"/>
      <xsd:element ref="dcdr:FrameworkCapabilities" minOccurs="0" maxOccurs="unbounded"/>
      <xsd:element ref="dcdr:NetworkLoad" minOccurs="0" maxOccurs="1"/>
  </xsd:sequence>
</xsd:complexType>
<xsd:complexType name="RackType">
      <xsd:sequence >
              <xsd:element  name="computedPower" type="dcdr:PowerType" minOccurs="0"
maxOccurs="1"/>
      <xsd:element ref="dcdr:RackableServer" minOccurs="0" maxOccurs="unbounded"/>
      <xsd:element ref="dcdr:Fan" minOccurs="0" maxOccurs="unbounded"/>
      <xsd:element ref="dcdr:WaterCooler" minOccurs="0" maxOccurs="unbounded"/>
      <xsd:element ref="dcdr:RackableRouter" minOccurs="0" maxOccurs="unbounded"/>
      <xsd:element ref="dcdr:RackableSwitch" minOccurs="0" maxOccurs="unbounded"/>
      <xsd:element ref="dcdr:SAN" minOccurs="0" maxOccurs="unbounded"/>
      <xsd:element ref="dcdr:Enclosure" minOccurs="0" maxOccurs="unbounded"/>
      <xsd:element ref="dcdr:PDU" minOccurs="0" maxOccurs="unbounded"/>
  </xsd:sequence>
</xsd:complexType>

<xsd:complexType name="ServerType">
      <xsd:sequence>
      <xsd:element  name="name" type="dcdr:ServerRoleType" minOccurs="1" maxOccurs="1"/>
      <xsd:element  name="status" type="dcdr:ServerStatusType" minOccurs="1" maxOccurs="1"/>
```

Fig. 5.12 Other example of complex type of entities of the meta-model in XSD format. As this example is for demonstration purposes, it may not show all the details

and hypervisors, and the comprehensive capture of energy-related parameters of the ICT resources, plays a crucial role in achieving the goals of the EMS. This approach is vital for reducing the energy consumption of the data center effectively.

5.4 Demand Adaptation

5.4.1 High-Level Overview

The primary goal of this EMS is to adjust the demand of data centers, whether by increasing or decreasing it. This falls under the umbrella of demand-response and demand-side management. This strategy has caught the interest of DCOs in recent years, as peak demand is becoming increasingly important in determining monthly energy tariffs, sometimes even more so than energy consumption itself. Furthermore, when there is an excess of energy from renewable sources, electricity prices may drop or even become negative.

This incentivizes data centers to increase their demand and utilize the energy surplus. It is important to distinguish this optimization from the one discussed in Sect. 5.3: The former (e.g., the approach of this section) focuses on adapting (e.g.,

This XML file does not appear to have any style information associated with it. The document tree is shown below.

```
<ns2:EMS xmlns:ns2="http://www.dcdr.com/schemes/MetaModel">
  <ns2:Site>
    <PUE>1.20</PUE>
    <CUE>1.20</CUE>
    <computedPower>288.21912736492783</computedPower>
    <ns2:Datacenter>
      <computingStyle>CLOUD</computingStyle>
      <computedPower>288.21912736492783</computedPower>
      <ns2:Rack>
      ...
      </ns2:Rack>
      <ns2:Rack>
      ...
      </ns2:Rack>
      <ns2:FrameworkCapabilities id="Nagios">
      ...
      </ns2:FrameworkCapabilities>
    </ns2:Datacenter>
  </ns2:Site>
</ns2:EMS>
```

Fig. 5.13 Illustrative of applying the template shown in Figs. 5.10, 5.11 and 5.12 is an instance featuring a single site with a PUE and CUE of 1.2. This site includes a cloud-type data center with a computed power of 288.2 kW. Additionally, the data center comprises two racks and incorporates a Nagios framework capability

reducing or increasing) the demand (in kW) during demand-response periods, which typically occur only limited times a year. In contrast, the latter aims to minimize energy consumption (in kWh) over the long term, emphasizing energy efficiency.

Figure 5.15 provides a high-level overview of the key actors of such an EMS. Particularly, the ecosystem for demand-response and demand-side management involves the following three distinct actors:

- Grid operator (GO): This actor is responsible for maintaining the stability of the grid infrastructure, ensuring its proper maintenance, and expanding it if necessary.
- Data center operator (DCO): This actor is responsible for ensuring the availability and reliability of the ICT infrastructure, managing and optimizing data center resources, planning for scalability and growth, complying with regulations and industry standards, and decreasing operational expenditure.
- IT customer: The ITC is the end-user of the services provided by a specific data center. They rely on the data center's ICT resources to run their services.

To ensure a certain level of service quality, the DCO and ITCs enter into service level agreements (SLAs). These agreements define the expectations and commitments regarding the quality of service (QoS), performance, availability, and other aspects of the provided services. Traditionally, these SLAs have strict constraints to guarantee the agreed-upon QoS. To avoid breaching the terms of the SLAs, the DCOs

This XML file does not appear to have any style information associated with it. The document tree is shown below.


```xml
<ns2:EMS xmlns:ns2="http://www.dcdr.com/schemes/MetaModel">
  <ns2:Site>
    <PUE>1.20</PUE>
    <CUE>1.20</CUE>
    <computedPower>288.21912736492783</computedPower>
    <ns2:Datacenter>
      <computingStyle>CLOUD</computingStyle>
      <computedPower>288.21912736492783</computedPower>
      <ns2:Rack>
        <computedPower>288.21912736492783</computedPower>
        <ns2:Enclosure frameworkRef="Nagios">
          <powerIdle>100.0</powerIdle>
          <powerMax>10000.0</powerMax>
          <frameworkID>Encl1</frameworkID>
          <computedPower>203.8976828175472</computedPower>
          <ns2:BladeServer frameworkRef="Nagios">
            <name>CLOUD_NODE_CONTROLLER</name>
            <status>ON</status>
            <frameworkID>cloudblade03</frameworkID>
            <computedPower>168.06196545278902</computedPower>
            <measuredPower>119.0</measuredPower>
            <ns2:Mainboard>
            ...
            </ns2:Mainboard>
            <ns2:NativeOperatingSystem>
              <name>LINUX</name>
              <ns2:HostedHypervisor>
                <hypervisorName>Xen</hypervisorName>
                <frameworkID>XEN13</frameworkID>
              </ns2:HostedHypervisor>
            </ns2:NativeOperatingSystem>
          </ns2:BladeServer>
          <ns2:PSU>
            <load>100.0</load>
            <continousPower>1000.0</continousPower>
            <peakPower>1200.0</peakPower>
            <efficiency>90.0</efficiency>
            <redundancy>0</redundancy>
            <computedPower>10.0</computedPower>
          </ns2:PSU>
          <ns2:PSU>
          ...
          </ns2:PSU>
          <ns2:Fan>
          ...
          </ns2:Fan>
          <ns2:Fan>
          ...
          </ns2:Fan>
          <ns2:Fan>
          ...
          </ns2:Fan>
          <ns2:Fan>
          ...
          </ns2:Fan>
          <ns2:Fan>
          ...
          </ns2:Fan>
          <ns2:Fan>
          ...
          </ns2:Fan>
        </ns2:Enclosure>
      </ns2:Rack>
      <ns2:Rack>
      ...
      </ns2:Rack>
      <ns2:FrameworkCapabilities id="Nagios">
      ...
      </ns2:FrameworkCapabilities>
    </ns2:Datacenter>
  </ns2:Site>
</ns2:EMS>
```


Fig. 5.14 A more comprehensive breakdown of the example depicted in Fig. 5.13 reveals that each of the two racks within the data center contains a single enclosure. This enclosure exhibits an idle and maximum demand of 100 kW and 10 MW, respectively, with a computed demand of 203.9 kW. Additionally, the enclosure includes one blade server, 3 power supply units (PSUs), and 6 fans

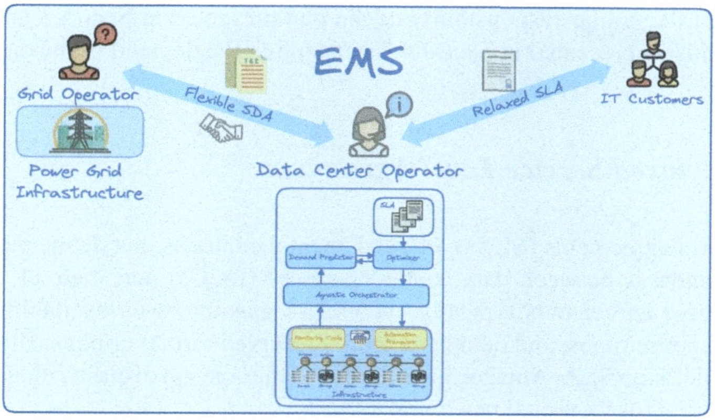

Fig. 5.15 A high-level overview of the key architectural components of an energy management system (EMS) designed to adapt or adjust the demand of a data center. In the ecosystem of such an EMS, three distinct actors are involved. The grid operator (GO) focuses on maintaining grid stability as its primary objective. The data center operator (DCO) aims to minimize operational costs while maximizing the usage of the data center (e.g., selling more services). The IT customer (ITC) utilizes the ICT resources provided by the data center to run its services. To facilitate their collaboration, the GO and DCO establish bilateral agreements known as supply-demand agreements (SDAs), which are electricity contracts. Similarly, the DCO and ITC have service level agreements (SLAs) to govern their relationship. Two pivotal concepts namely the "Relaxed SLA" and the "Flexible SDA" are necessary to enable the demand adaption collaboration between the three involved actors of the ecosystem

often over-provision their ICT resources. This over-provisioning leads to increased power demand, energy consumption, and ultimately higher operation costs.

In contrast, the grid operator (GO) and its energy consumers (in this book data center operators—DCOs), enter into an electricity contract known as a supply-demand agreement (SDA). For instance, in Germany and for big industrial consumers like data centers, it outlines (1) fees related to the usage of the grid infrastructure as well as the highest power requested during the billing period, (2) an electricity price (such as time-of-use—ToU) for each kilowatt-hour consumed, and (3) different taxes.

These terms remain unchanged throughout the year, regardless of the operational state of the power grid. This lack of flexibility in the contract leads to a complete absence of adaptability in the power supply. Additionally, the existing contracts fail to consider the need for consumers to increase their power demand when there is an excess of energy generated by renewable sources. Therefore, both service level agreements (SLAs) and supply-demand agreements (SDAs) are not appropriate for implementing demand adaptation in light of the aforementioned limitations.

Consequently, it is necessary to relax these agreements from a contractual standpoint to leverage the flexibilities offered by the consumers for effective demand adaptation and demand-side management. It is important to note that the EMS inside the data center eco-system is responsible for managing the ICT resources of the data center. It has a similar responsibility as the one presented in Sect. 5.3 and exploits the mechanisms presented in Sect. 4.4.2 in adapting the demand of the data center.

5.4.2 *Relaxed Service Level Agreement*

Service level agreements (SLAs) are fundamental contracts that define the expected service standards between data center operators (DCOs) and their IT customers (ITCs). These agreements typically encompass guarantees on availability, performance, response times, and details regarding reserved virtual or physical resources. For example, as an SLA, Amazon has introduced the concept of elastic cloud compute units (EC2Us) for its virtual machines.

An EC2U represents the compute units allocated to each instance multiplied by the number of virtual cores, facilitating comparison of capabilities across different Amazon EC2 instance types. Amazon offers a diverse range of over 250 instance types, categorized into "general purpose", "compute optimized", "memory optimized", "accelerated computing," "storage optimized," and "HPC optimized". Each category includes pre-configured instances specifying the amount of resources for computation (vCPU), storage, memory, and network bandwidth (Amazon 2024).

Table 5.1 showcases 10 instance types from the "general purpose" category, specifically powered by ARM-based processors. All of those instances have an SLA stating 99.99% of availability providing a secure platform for computation, communication, and storage purposes. However, as previously mentioned, traditional SLAs

Table 5.1 This table presents an overview of 10 distinct instance types available in Amazon's elastic cloud computing unit (EC2U) for the "general purpose" category. The presented instance types are powered by ARM-based AWS Graviton3 processors. Amazon offers an extensive selection of over 250 different instance types within the EC2U, ensuring compatibility with diverse computing needs and requirements (Amazon 2024)

Instance size	vCPU	Memory (GiB)	Storage (GiB)	Network bandwidth (Gbps)
m7g.medium	1	4	EBS-Only	Up to 12.5
m7g.large	2	8	EBS-Only	Up to 12.5
m7g.xlarge	4	16	EBS-Only	Up to 12.5
m7g.2xlarge	8	32	EBS-Only	Up to 15
m7g.16xlarge	64	256	1 × EBS-Only	30
m7gd.medium	1	4	1 × 59 NVMe SSD	Up to 12.5
m7gd.large	2	8	1 × 118 NVMe SSD	Up to 12.5
m7gd.large	2	8	1 × 118 NVMe SSD	Up to 12.5
m7gd.xlarge	4	16	1 × 237 NVMe SSD	Up to 12.5
m7gd.16xlarge	64	256	2 × 1900 NVMe SSD	30

often come with rigid parameters that can limit their applicability in the context of demand-side management.

Relaxed SLAs, on the other hand, introduce a more adaptable framework that encourages flexibility and innovation within the data center industry suitable for demand-side management (DSM) (Amokrane et al. 2015; Basmadjian et al. 2013; Haque et al. 2013; Hasan et al. 2015; Wang et al. 2013). More precisely, relaxed SLAs redefine the relationship between the DCO and ITCs by offering a more dynamic approach.

They relax the performance parameters of the traditional SLAs, incorporate new performance parameters relevant to demand-side management, and foster collaboration through incentives. Relaxed SLAs are not just about easing restrictions; they are about reimagining how data centers in collaboration with their ITCs can operate with greater flexibility necessary for participating in demand adaptation and demand-side management.

This concept requires the DCOs and their ITCs to work together. In this regard, the ITCs agree to some variability in the promised service quality for specific periods in exchange for incentives like cheaper service prices. This variability is based on the data center's needs concerning demand adaptation. When there's no need for changes, the ITCs get the promised service quality.

If the data center doesn't meet the agreement without a valid reason, there are penalties (as is the case with the traditional SLAs). However, when demand-side management is required by the data center to adapt (e.g., reduce) power, then based on the amount of demand to be adapted, the ITCs through relaxed SLA accept medium (e.g., 80% of the original performance) to high (e.g., 65% of the original performance) degradation in the performance of the services, in return for monetary incentives.

In general, relaxed SLAs can be structured based on the required flexibility, taking into account either a calendar or operation mode context. In the case of a calendar-based context, the agreed level of performance can be determined based on specific hours and days of the week. For example, it could be agreed upon that the system operates at (1) full performance on weekdays between 8 AM and 8 PM, medium performance between 8 PM and 8 AM, and (2) medium performance during weekends. Alternatively, the relaxed SLAs can be structured based on the mode of operation of the data center. Typically, three modes are defined: regular, power-saving, and power-spending.

By default, the data center operates in regular mode. However, upon receiving a request from its GO to either reduce or increase power demand, the data center may switch to power-saving or power-spending modes for a certain period respectively. In the power-saving mode, the data center aims to reduce demand and hence save energy by downgrading the performance of the IT services, migrating or consolidating virtual machines (VMs), and shifting tasks in time. Various mechanisms, as described in Sect. 4.4.2.1, can be employed to achieve these objectives.

On the other hand, in the power-spending mode, the data center promotes IT services to higher performance levels, anticipates maintenance tasks, cools down the facility (as explained in Sect. 4.4.2.2), and charges the uninterruptible power supply

(UPS) (as described in Sect. 4.4.2.3). These actions are taken to utilize excess energy from renewable sources. Overall, the flexibility of relaxed SLAs can be tailored to accommodate specific calendar-based requirements or adapt to different operational modes of the data center, ensuring efficient resource utilization and meeting varying performance needs.

5.4.3 Flexible Supply-Demand Agreement

Currently, there is a lack of direct feedback between the GO experiencing some pressure on the grid and the energy consumers contributing to this pressure. The pressure arises from either an uncontrollable and volatile power supply or unexpected demand spikes from major energy consumers such as data centers.

In the past, power shortage situations were addressed through direct load control (DLC), where the GO would communicate with each significant energy consumer (e.g., industrial facilities) to regulate and control their equipment. However, replicating this approach in the data center industry is impractical due to the complex nature of data centers and their processes. Specifically, it is not feasible for the GO to directly control and manage every individual server in a data center.

Therefore, it is necessary to transfer the responsibility of managing the data center infrastructure for demand-side management purposes to the DCO. However, this transfer requires a bilateral agreement between the GO and DCO, outlining the extent and nature of collaboration between the two parties. The presented agreement is characterized as a flexible supply-demand agreement (SDA), emphasizing its ability to offer flexibility for both parties involved in practicing demand adaption and demand-side management.

This book delves into the concept of a flexible SDA, which is introduced as an extension to existing energy contracts. The exploration of this concept is based on its origin from Basmadjian et al. (2016). This agreement outlines the conditions under which the GO can request demand adaptation actions from its consumers. While the concept itself applies to various industries, the book focuses on a domain-specific instantiation for the data center industry. This instantiation takes into account the unique characteristics of data centers, which are known for their high flexibility and responsiveness.

To promote collaboration between the GO and data centers for demand adaptation, it is necessary to make changes to the existing energy contracts. As mentioned previously, the primary motivation behind introducing new types of energy contracts is the limited flexibility of the current contracts in terms of power adaptation and demand-side management.

To address this, the flexible SDA outlined in this book includes contractual terms that facilitate collaboration between the GO and its consumers (e.g., data centers) for demand adaptation purposes. Please be aware that the contractual terms mentioned can serve as a template for future agreements between the GO and various types of consumers, not just limited to data centers as discussed in this book. In the rest of this section, we use consumers to denote data centers.

To begin, the contract should include specific terms regarding demand adaptation, outlining the minimum and maximum amount of demand that consumers can increase and/or decrease in response to factors like excess of renewable energy or sudden peak demand. Each of these scenarios should also specify the minimum and maximum duration for which consumers can commit to these actions, such as a minimum increase or maximum decrease. The combination of demand adaptation as well as duration information constitutes the so-called demand adaptation profile.

In Fig. 5.16, there are four different profiles shown that represent the range of durations and demand adaptation options. For example, the green point indicates that the consumer can adjust their power usage by 0.5 MW for a duration of 1 h. These types of adaptation profiles are not ideal for demand-side management (DSM) as they offer very limited flexibility.

On the other hand, the red rectangle represents a highly flexible and suitable profile for DSM. It allows for demand adaptations ranging from 1 to 3 MW, with durations ranging from 1 to 4 h. It is important to note that demand adaptation can both increase and decrease the demand, so it is possible to have one profile for reducing demand and another for increasing power usage (see Fig. 5.17).

The collaboration between the GO and consumers should establish the periods during which the GO can send demand adaptation requests. This can be defined as

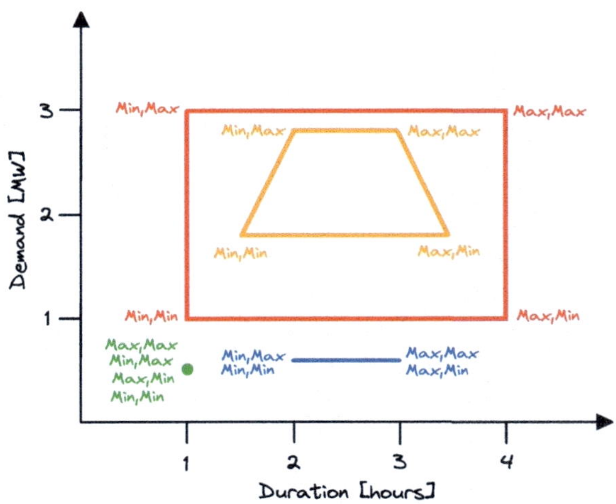

Fig. 5.16 The example outlines four distinct demand adaptation profiles based on duration and demand levels. The red rectangle illustrates a range from 1 MW to 3 MW in demand adaptation, with durations of 1–4 h. The orange trapezoid shows a minimum adaptation of 1.8 MW over 1.5–3.5 h. A maximum of 2.8 MW can be achieved within a duration of 2–3 h. The blue line depicts a profile with a constant adaptation of 0.6 MW, lasting 2–3 h. Finally, the green point highlights a special case with both minimum and maximum demand adaptations at 0.5 MW for a duration of 1 h, showcasing the least flexible profile among the four

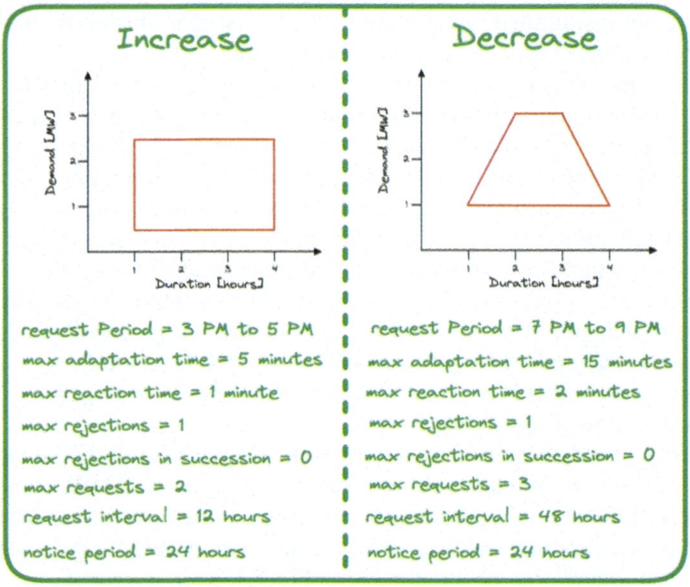

Fig. 5.17 The proposed concept highlights its customizability through a demand adaptation profile tailored for each specific month, illustrating its flexibility to meet individual needs. For instance, the example provided showcases an adaptation profile specifically designed for January

specific timeframes, such as morning, noon, afternoon, or night, or even more precise periods like 8 AM to 10 AM and 8 PM to 11 PM. For effective planning on the GO's side, the agreement should also specify the maximum time consumers require to adapt their demand, as well as the maximum time (usually not exceeding 5 min) for consumers to respond to demand adaptation requests of the GO.

To prevent manipulation from either party, it is important to define the maximum number of demand adaptation requests that the GO can send to consumers, as well as the maximum number of times consumers are allowed to reject the GO's requests. Additionally, there should be a limit on the number of successive rejections permitted by consumers. Since recovery from a demand adaptation needs to be done within a reasonable timeframe, the contract should establish a minimum period of time (in hours) between two successive demand adaptation requests from the GO to consumers.

This ensures that consumers who have already executed a demand adaptation request have enough time to recover before potentially rejecting a GO request and exceeding the agreed maximum number of rejections. Lastly, there should be guarantees that consumers have sufficient time (in hours) to receive notice from the GO regarding potential demand adaptation requests. This allows them to adequately prepare and respond to requests promptly.

To monitor the actual execution of contractual terms by the consumers, three specific terms need to be tracked on a monthly basis: the maximum (1) number of rejections, (2) number of successive rejections, and (3) number of requests. These terms are stored as separate pieces of information, such as variables, and they need to be reset at the start of each month. The modularity of the mentioned contractual terms enables highly adaptable flexible SDAs to create various demand adaptation profiles for different situations.

For example, a grid operator can create for each consumer unique adaptation profiles for each month, allowing for customization based on specific needs. Furthermore, the demand adaptation profile of a consumer for increasing demand in January can be completely distinct from that for decreasing demand. These detailed demand adaptation profiles provide the demand profile selector algorithms (also known as algorithms for allocation strategies) of the GO with the ability to adapt to various DSM scenarios.

Figure 5.17 and 5.18 show different demand adaptation profiles for power increase and decrease with different contractual terms for the months of January and August respectively. Similar demand adaptation profiles can be created between the GO and its consumers for the other months of the year. Whenever a party violates any terms of the flexible SDA, it is necessary to impose a penalty.

Additionally, the agreed-upon contractual terms are linked to a financial incentive. This ensures that the most adaptable consumers receive the greatest reward, as they assist in the planning of the GO by participating in demand-side management. The rewards and penalties in this book are based on extensively researched economic schemes, but the details of these schemes are beyond the scope of this book. However, for illustrative purposes, we will now present some basic reward and penalty schemes.

Reward

The reward of the consumers can be composed of two parts such as static and dynamic as:

$$\text{Reward}_{\text{Consumer}} = \text{Reward}_{\text{Static}} + \text{Reward}_{\text{Dynamic}} \tag{5.1}$$

The static reward remains consistent throughout the duration of a signed flexible SDA, while the dynamic reward varies based on the level of consumers' adherence to their promised demand adaptation. To design static rewards, a straightforward approach is to establish three flexibility categories: small, medium, and large.

Once these categories are defined, specific rules need to be outlined to determine which agreement terms fall into each category. For example, agreement terms related to demand adaptation can be categorized such that adaptations below 1 MW are considered small, adaptations between 1 and 10 MW are medium, and adaptations exceeding 10 MW are large.

Similarly, for terms related to the duration of adaptation, the same categories can be assigned, with durations less than 1 h classified as small, durations between 1 and 3 h as medium, and durations exceeding 3 h as large. It is important to note that the numerical values provided in this example are for illustrative purposes only and

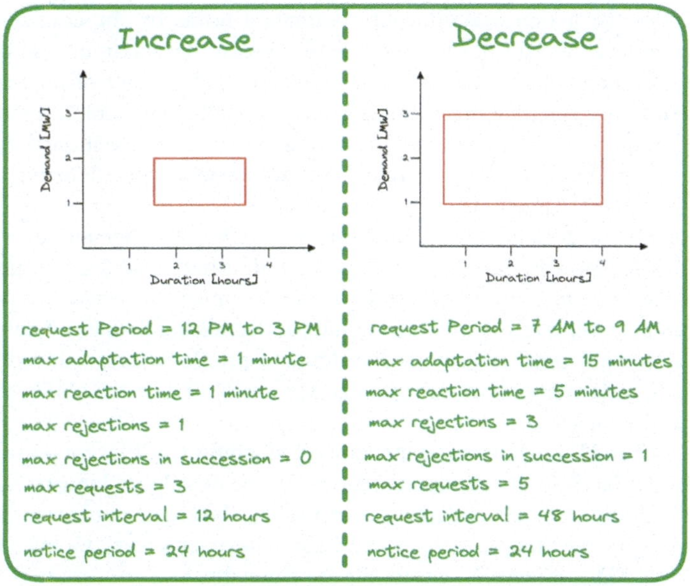

Fig. 5.18 The proposed concept emphasizes its customizability by offering a demand adaptation profile uniquely crafted for each month, demonstrating its ability to cater to individual preferences. For example, the profile presented is specifically tailored for August

should be adjusted based on the specific flexibilities and capabilities of the consumers involved.

When considering the term related to the maximum time for adaptation, the most flexible consumers are those who require less time. If the maximum adaptation time is below 10 min, it falls into the large group. If it falls between 10 and 15 min, it belongs to the medium group. Any maximum adaptation time exceeding 15 min is assigned to the small group.

In similar fashion, the groups can also be assigned based on the maximum time consumers need to react to the GO's request for demand adaptation. If the reaction time is less than 1 min, it belongs to the large group. If the reaction time falls between 1 and 3 min, it is assigned to the medium group. Otherwise, it belongs to the small group.

Concerning the terms of maximum rejections and rejections in succession, consumers are considered more flexible when these values are lower. If the maximum number of rejections is below 2, it is classified as large. If it falls between 2 and 3, it belongs to the medium group. Otherwise, it is classified as small. Similarly, for the term related to the maximum number of rejections in succession, a value of zero is assigned to the large group. If it falls between 1 and 2, it belongs to the medium group. Any other value is classified as small.

Referring to the term that specifies the maximum requests that the GO can send to its consumers if the value is less than 2, it belongs to the small group. If it falls between 2 and 4, it is classified as medium. And if it exceeds 4, it is considered large. Lastly, the term that determines the interval in which the GO must wait before sending another adaptation request to consumers also plays a role in defining flexibility. Specifically, if this interval is less than 13 h, it falls into the large group. If it falls between 13 and 24 h, it is categorized as medium. Any interval exceeding 24 h is classified as small.

Similarly, when considering the term for the notification period, if it is less than 13 h, it is classified as large. If it falls between 13 and 24 h, it belongs to the medium group. Any other value is categorized as small. After assigning each contractual term of the flexible SDA to one of the mentioned groups, the static reward can be determined. Specifically, the numerical values 1, 2, and 3 correspond to the categories of small, medium, and large respectively.

With a total of 26 terms (13 for each adaptation possibility, such as increase or decrease), the most flexible consumer will have a score of 78 (e.g., 26 multiplied by 3). Note that if a contractual term is left empty, it will be assigned a value of zero. Then the static reward can be computed as:

$$\text{Reward}_{\text{Static}} = \text{Fee}_{\text{Demand}} \cdot \frac{39}{\sum_{i=1}^{26} \text{Score}_i} \tag{5.2}$$

where $\text{Fee}_{\text{Demand}}$ refers to the basic fee that consumers pay for the highest demand specified in their electricity contract. On the other hand, Score_i represents the value assigned to each individual contractual term i, which is determined based on the group it is categorized into. Therefore, the most flexible consumers who are adapting their power usage by both increasing and decreasing it, and fall into the large category for all 26 terms, receive a 50% discount on the basic fee they pay for their power demand.

In the given example of the flexible SDA shown in Fig. 5.17, when we assign the agreed-upon contractual terms to their respective groups and scores, we can calculate that the corresponding consumer receives a score of 31 for power increase and 28 for power decrease adaptations. As a result, the corresponding consumer is eligible for a 33.9% discount (calculated as $1 - \frac{39}{31+28}$) on the basic fee for their power demand.

The flexible SDA grants a dynamic reward to each consumer at the end of a billing period (e.g., monthly) for their responsiveness to specific demand adaptation requests from the GO. This reward reflects the actual grid support provided by the consumers. To estimate the level of support provided, we need to first calculate the collaboration level using the following key information:

- The requested amount $\text{P}_{\text{Requested}}$ and duration $\text{T}_{\text{Requested}}$ of demand adaptation by the GO.
- The actual amount $\text{P}_{\text{Adapted}}$ and duration $\text{T}_{\text{Adapted}}$ of demand adaptation achieved by the consumer.

- The total number of demand adaptation requests $N_{Requested}$ sent by the GO to the underlying consumer.
- The number of rejections $N_{Rejected}$ of the GO's requests by the consumer.

These information allow us to compute the monthly collaboration level as such:

$$Collab_{Consumer} = \frac{P_{Adapted} \cdot T_{Adapted}}{P_{Requested} \cdot T_{Requested}} \cdot \frac{N_{Requested} - N_{Rejected}}{N_{Requested}} \tag{5.3}$$

It is worthwhile to note that $Collab_{Consumer} \in [0, 1]$, where a value of 1 indicates the highest level of collaboration between the two parties, and 0 denotes otherwise.

Once the level of collaboration is specified, the next step consists of computing the dynamic reward as:

$$Reward_{Dynamic} = P_{Adapted} \cdot T_{Adapted} \cdot Collab_{Consumer} \cdot Price^{DSM}_{Energy} \tag{5.4}$$

where $Price^{DSM}_{Energy}$ refers to the price of electricity paid by the GO to its consumer for the adapted energy consumption, measured in kilowatt-hours (kWh).

It is important to set this price significantly higher than the regular electricity prices to incentivize consumers to actively participate in demand-side management (DSM) initiatives. The actual value for $Price^{DSM}_{Energy}$ is determined through negotiations when signing the flexible SDA and can vary between different GOs. However, a reasonable approach could be to set the $Price^{DSM}_{Energy}$ a few factors higher than the cost of electricity during peak hours. This ensures that consumers have a strong incentive to modify their demand usage patterns and contribute to the overall demand adaptation and DSM.

To calculate the dynamic reward using Eq. (5.4), let's examine the scenario provided. We have a data center with 3000 rack servers, each having a demand of 3 kW. Therefore, the maximum demand required by the data center's ICT infrastructure is approximately 9 MW. Additionally, the cooling system of the data center has a demand of 1 MW, resulting in a maximum overall demand of 10 MW. Assuming that the average demand of the data center is around 3 MW, the total monthly energy consumption of the data center would be approximately 2232 MWh (3 MW × 24 h × 31).

Considering the average electricity prices for non-household consumers, such as data centers, which is 21 cents/kWh in Europe (based on 2023 statistics), the monthly electricity cost for this data center would be around €0.468 M. Furthermore, in this scenario, the data center is a perfect collaborator with the GO, indicated by $Collab_{Consumer} = 1$. Referring to the flexible SDA for the month of January in Fig. 5.17, let's assume that the GO sent two power decrease requests, each requesting a reduction of 3 MW for 1 h.

As the data center is a perfect collaborator, it accepts both requests and adapts a total of 6 MWh. Based on Eq. (5.4), the dynamic reward that the data center will receive for its collaboration is calculated to be about €50k assuming that $Price^{DSM}_{Energy}$ is agreed to be about €8.33/kWh. Therefore, in the aforementioned scenario where consumers collaborate perfectly and utilize the flexible SDA shown in Fig. 5.17, it can be inferred that they are eligible for a 34% (calculated above) discount on the costs associated with the monthly tariff for the highest demand. Additionally, they can enjoy an extra reduction of approximately 11% (e.g., $\frac{50k}{468k}$) in the costs linked to energy consumption by fulfilling two demand adaptation requests from the GO.

Penalty

This scheme is put into action when consumers do not uphold their promises to the GO during demand-side management events, such as demand adaptation, by breaking one or more contractual terms of the flexible SDA. As a result, this scheme should not penalize consumers who have demonstrated their cooperation through demand adaptation (e.g., $\text{Collab}_{\text{Consumer}} \sim 1$) in the same way as those who have committed but shown less or no collaboration. Similar to the approach we took for the reward scheme, we suggest implementing a straightforward penalty scheme:

$$\text{Penalty}_{\text{Consumer}} = \text{P}_{\text{Requested}} \cdot \text{T}_{\text{Requested}} \cdot \text{Price}_{\text{Energy}}^{\text{Peak}} \cdot (1 - \text{Collab}_{\text{Consumer}}) \qquad (5.5)$$

where $\text{P}_{\text{Requested}}$, $\text{T}_{\text{Requested}}$, and $\text{Collab}_{\text{Consumer}}$ are defined in Eq. (5.3), whereas $\text{Price}_{\text{Energy}}^{\text{Peak}}$ denotes the energy price during peak hours.

Therefore, if we assume that the data center mentioned earlier has a collaboration level of 50% and the same scenario for demand adaptation as above, and considering a peak energy price of €2.0825/kWh agreed upon in the flexible SDA, a penalty of €6247.5 will be enforced. It is important to note that Eq. (5.5) ensures that consumers who collaborate less will face higher penalties. Furthermore, the negotiation of both $\text{Price}_{\text{Energy}}^{\text{DSM}}$ and $\text{Price}_{\text{Energy}}^{\text{Peak}}$ is necessary during the signing of the flexible SDA, and the prices provided in the above examples are solely for demonstration purposes.

The proposed reward and penalty schemes are specifically designed to ensure that consumers with a collaboration level of 50% are still motivated to participate in demand-side management, as there will always be a positive value remaining despite the penalties they may have to pay. According to Eq. (5.3), it can be argued that a collaboration level of 50% can be achieved by either (1) rejecting one of the two adaptation requests from the GO and performing the adaptation in one shot, or (2) by accepting both requests but only adapting half of the requested demand.

Interestingly, for the former case, Eq. (5.3) provides much higher rewards compared to the latter case. However, this scenario is prevented because consumers do not have prior knowledge of the amount of demand adaptation required, as well as the duration and exact timing of the requested adaptation.

Therefore, even if they intentionally reject requests, their collaboration level will decrease. This mechanism is designed to incentivize consumers to truthfully collaborate with the GO. Furthermore, it is possible to develop more complex business models for reward and penalty schemes, as well as design mechanisms. However, the details of such models and mechanism designs are beyond the scope of this chapter and book.

5.5 Summary

Human-centered cyber-physical systems (HCPS) such as data centers require a clearly defined feedback control loop process to enable optimizations like energy efficiency and demand-side management. This chapter initiated by outlining a comprehensive feedback control loop process, which commences with evaluating the HCPS environment concerning hardware, operators, and stakeholders. The loop concludes by implementing a series of optimal decisions to reconfigure the infrastructure through well-crafted human-machine interfaces (HMI).

Subsequently, the chapter introduces the software-based orchestration of the energy management system (EMS). Initially, all essential components of an EMS are outlined, including data acquisition, analysis, monitoring and reporting, optimizations, remote control and automation, as well as actuation and integration. The importance of an agent-based approach in implementing EMS within data centers is then underscored.

Specifically, it is demonstrated that the multi-agent concept goes beyond mere abstraction due to the autonomous nature of agents. It is further explained that this concept comprises two key elements: allocation strategies and protocol. Allocation strategies define the rules for agent interaction, while the protocol facilitates information exchange among agents. In this context, the contract net protocol (CNP) is introduced, and its components are elucidated.

In the third section of this chapter, the focus was on an EMS designed to reduce a data center's energy consumption while meeting the SLAs of its IT customers. The section began by providing a broad overview of the key architectural components of such an EMS, highlighting the importance of the "Demand Predictor" and "Optimizer".

Additionally, it was illustrated that through the use of connectors, the orchestrator can operate independently of the diverse hardware and software within the system, making it agnostic. The section was wrapped up by introducing a generic XSD template for describing the ICT resources within a data center, which was further exemplified for better understanding. The corresponding instance serves as an input to the optimization engine of the underlying EMS.

The last section of the chapter deals with an EMS that is used for demand adaptation (e.g., increase or decrease) of an ecosystem consisting of three different actors. Thus, the involved actors within the ecosystem of the grid operator (GO), data center operator (DCO), and IT customers (ITC) were first presented, showing the bilateral agreements between them. It demonstrated that the current rigid SLAs and SDAs are not suitable for demand adaptation and DSM. To address this issue, the concept of relaxed SLAs based on the calendar or mode of operation context were introduced, accompanied by illustrative examples.

Furthermore, flexible SDAs were presented, along with the presentation of relevant contractual terms. The section has also suggested simple yet effective reward and penalty schemes, and a real-world flexible SDA has been instantiated as an exemplar. Importantly, it was shown that even with a 50% collaboration rate between DCOs and the GO, there are still incentives for DCOs to actively participate in demand-response activities, such as demand adaptation and DSM. Overall, this section highlights the importance of flexible agreements in enabling effective DSM practices and encourages further exploration and implementation of such approaches in the future.

References

Amazon. Amazon ec2. https://aws.amazon.com/ec2/instance-types/. Accessed 05 Feb 2024

Amokrane A, Langar R, Zhani MF, Boutaba R, Pujolle G (2015) Greenslater: on satisfying green slas in distributed clouds. IEEE Trans Netw Serv Manag 12(3):363–376

Barroso LA, Clidaras J (2022) The datacenter as a computer: an introduction to the design of warehouse-scale machines. Springer Nature

Barth W (2008) Nagios: system and network monitoring. No Starch Press

Basmadjian R, Botero JF, Giuliani G, Hesselbach X, Klingert S, De Meer H (2016) Making data centers fit for demand response: Introducing greensda and greensla contracts. IEEE Trans Smart Grid 9(4):3453–3464

Basmadjian R, Lovasz G, Beck M, De Meer H, Hesselbach-Serra X, Botero JF, Klingert S, Ortega MP, Lopez JC, Stam A et al (2013) A generic architecture for demand response: the all4green approach. In: 2013 international conference on cloud and green computing, pp 464–471. IEEE

Basmadjian R, Meer HD, Lent R, Giuliani G (2012) Cloud computing and its interest in saving energy: the use case of a private cloud. J Cloud Comput: Adv Syst Appl 1:1–25

Benson RM, Munsell E, Bertrand N, Baynton M, Bollig EF, McDonald J (2019) A multi-environment hpc-scale puppet infrastructure for compliance and systems automation. In: Proceedings of the practice and experience in advanced research computing on rise of the machines (learning), pp 1–8

Coffman EG, Garey MR, Johnson DS (1984) Approximation algorithms for bin-packing-an updated survey. In: Algorithm design for computer system design, pp 49–106. Springer

Dalle Vacche A (2015) Mastering Zabbix. Packt Publishing Ltd

Fekete SP, Schepers J (2001) New classes of fast lower bounds for bin packing problems. Math Progr 91:11–31

Frank F, Alfke M, Franceschi A, Pastor JS, Uphillis T (2017) Puppet: mastering infrastructure automation. Packt Publishing Ltd

Garey MR, Johnson DS (1981) Approximation algorithms for bin packing problems: a survey. In: Analysis and design of algorithms in combinatorial optimization, pp 147–172. Springer

Guccione S, McKirahan J (2016) Human machine interface: concepts and projects. G - Reference, information and interdisciplinary subjects series. Industrial Press, Incorporated

Hall J (2016) Mastering SaltStack. Packt Publishing Ltd

Haque ME, Le K, Goiri Í, Bianchini R, Nguyen TD (2013) Providing green slas in high performance computing clouds. In: 2013 international green computing conference proceedings, pp 1–11. IEEE

Hartson HR, Hix D (1989) Human-computer interface development: concepts and systems for its management. ACM Comput Surv (CSUR) 21(1):5–92

Hasan MS, Kouki Y, Ledoux T, Pazat J-L (2015) Exploiting renewable sources: When green sla becomes a possible reality in cloud computing. IEEE Trans Cloud Comput 5(2):249–262

Hochstein L, Moser R (2017) Ansible: up and running: automating configuration management and deployment the easy way. O'Reilly Media, Inc

Josephsen D (2007) Building a monitoring infrastructure with Nagios. Prentice Hall PTR

Keshav S (2012) Mathematical foundations of computer networking. Addison-Wesley

Koomey J, Brill K, Turner P, Stanley J, Taylor B (2007) A simple model for determining true total cost of ownership for data centers. Uptime Institute White Paper, Version 2, p 279

Myers BA (1998) A brief history of human-computer interaction technology. Interactions 5(2):44–54
Olups R (2010) Zabbix 1.8 network monitoring. Packt Publishing Ltd
Seiden SS (2002) On the online bin packing problem. J ACM (JACM) 49(5):640–671
Smith RG (1980) The contract net protocol: high-level communication and control in a distributed problem solver. IEEE Trans Comput 29(12):1104–1113
Taylor M, Vargo S (2014) Learning chef: a guide to configuration management and automation. O'Reilly Media, Inc
Turnbull J (2018) Monitoring with prometheus. Turnbull Press
Wang L, Khan SU, Chen D, Kołodziej J, Ranjan R, Xu C-Z, Zomaya A (2013) Energy-aware parallel task scheduling in a cluster. Future Gener Comput Syst 29(7):1661–1670
Wooldridge M (2009) An introduction to multiagent systems. Wiley

Part III
Methodologies for Power Demand Estimation and Power-Performance Trade-Off Analysis

The third part of this book introduces methodologies (1) to derive power prediction models for servers in data centers and (2) to analyze the power-performance trade-off as improving the performance of the server and decreasing its power demand and energy consumption are competing objectives. Chapter 6 emphasizes the significance of CMOS technology in microprocessors and data centers, tracing its evolution since the late 1960s and highlighting key advancements that improved power efficiency and performance. It introduces two analytical models for estimating power demand in CMOS circuits during idle and active states. Also, this chapter covers the von Neumann architecture, detailing its design and key components, such as the CPU and RAM, which are essential for understanding power-demand models. The final part focuses on power demand in data centers, identifying servers as major contributors to power demand and energy consumption. It presents methodologies for predictive models of idle and active power demands, detailing critical variables like voltage and frequency. This knowledge is vital for optimizing energy use and server performance in data centers, paving the way for advanced modeling techniques especially when considering the dynamic voltage and frequency scaling (DVFS) technology for the servers.

Chapter 7 examines how hardware manufacturers manage the balance between power demand and performance using dynamic voltage and frequency scaling (DVFS) technology. It begins with an in-depth overview of DVFS and its various governors, detailing their specific rules for adjusting processor frequencies and providing mathematical metrics for evaluating performance and power needs. The chapter progresses to queuing theory as a method for analyzing computing performance, discussing different system types and key metrics. It then moves to Markovian analysis, exploring its role in understanding the power-performance trade-off through Markov processes and chains. The appropriateness of the single-server M/M/1 queuing model for assessing this trade-off is established, with validation against real-world data. Lastly, the chapter introduces discrete-time Markov chain models for utilization-based governors, focusing on Demand and Conservative, and discusses steady-state probabilities to deepen the reader's grasp of these governance strategies.

Chapter 6
Analytical Power Prediction Models

6.1 CMOS Technology

6.1.1 History and Evolution

Complementary metal-oxide-semiconductor (CMOS) technology has a rich history that dates back to the late 1960s. It was developed as a way to address the limitations of that time's existing semiconductor technologies and has since become the dominant technology in the design of integrated circuits. The development of this technology can be traced back to the work of Chih-Tang Sah and Frank Wanlass at Fairchild Semiconductor in the late 1960s (Chih-Tang 1988). They proposed the concept of using both p-type and n-type transistors in a complementary configuration to reduce power demand and improve noise immunity in digital circuits.

In the early 1970s, researchers at RCA and IBM further advanced CMOS technology, demonstrating its potential for high-speed and low-power applications. By the mid-1970s, CMOS had become a key technology for digital integrated circuits, offering significant advantages over other technologies such as NMOS and PMOS. Throughout the 1980s and 1990s, CMOS technology continued to evolve (Bhuyan 2017; Shahidi 2007), focusing more specifically on improving performance, reducing power demand, and increasing integration levels.

One key development in CMOS technology has been the transition to smaller feature sizes, enabled by advancements in semiconductor manufacturing processes. For instance, the introduction of deep submicron CMOS processes in the 1990s enabled the development of highly complex and energy-efficient integrated circuits (Gu and Elmasry 1996; Lin et al. 2002), paving the way for the modern era of digital electronics.

Shrinking the size of transistors and other components on a chip allows for higher levels of integration, increased speed, and reduced power demand. This trend towards miniaturization has been driven by Moore's Law, which states that the number of

R. Basmadjian, *Demand-Side Energy and Power Management in Data Centers*,
Lecture Notes in Energy 1, https://doi.org/10.1007/978-3-031-85956-4_6

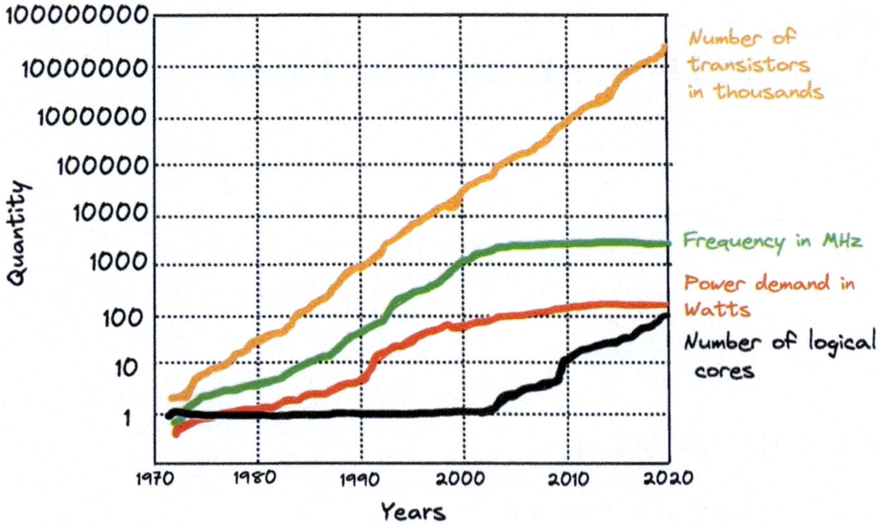

Fig. 6.1 An illustrative graph depicting Moore's Law from 1970 to 2020 reveals a noticeable level of saturation in the last two decades concerning frequency and power demand. During this time frame, CPUs have reached a maximum frequency of approximately 4.5 GHz. To counterbalance this trend, there has been a linear increase in the number of cores and transistors (nearly doubling) per CPU. By 2030, it is anticipated that microprocessors will feature over 1 trillion transistors within a single package! This advancement has significantly transformed the approach to parallel computations in data centers

transistors on a chip doubles approximately every two years (Schaller 1997) (see Fig. 6.1). As a result, modern CMOS integrated circuits can contain billions of transistors, enabling the development of complex systems-on-chip (SoCs) that combine multiple functions on a single die.

In addition to scaling down feature sizes, researchers have also focused on improving the performance of CMOS technology through innovations in device design, materials, and process technology. For example, the introduction of high-k dielectrics and metal gate electrodes has helped to reduce leakage currents and improve transistor performance, leading to higher speeds and lower power demand.

Fin field-effect transistor (FinFET) technology has been proposed, which offers improved electrostatic control and reduced leakage currents compared to traditional planar CMOS transistors (Jurczak et al. 2009). FinFETs feature a three-dimensional fin structure that helps to control the flow of current more effectively, enabling higher performance and lower power demand in integrated circuits.

Other advancements, such as the integration of advanced packaging technologies such as 3D integration and chip stacking and the development of novel device structures, have further enhanced the capabilities of CMOS technology. These approaches allow for the vertical integration of multiple chip layers, enabling higher levels of integration and improved communication between different functional blocks on a chip. By stacking memory, logic, and other components vertically, designers can reduce interconnect lengths, minimize signal delays, and improve overall system performance.

Additionally, recently researchers have been exploring the potential of emerging technologies, such as quantum computing and neuromorphic computing, to complement traditional CMOS technology. Quantum computing promises to revolutionize computation by leveraging the principles of quantum mechanics to perform complex calculations at speeds far beyond the capabilities of classical computers (Arute et al. 2019).

Neuromorphic computing, inspired by the structure and function of the human brain, aims to develop energy-efficient computing systems that can learn and adapt to new tasks (Marković et al. 2020). By integrating these emerging technologies with CMOS circuits, researchers hope to unlock new possibilities for high-performance computing, artificial intelligence, and other advanced applications.

Today, CMOS technology is ubiquitous in a wide range of applications, from microprocessors and memory chips to sensors and communication devices. Its scalability, low power demand, and compatibility with modern manufacturing processes have solidified its position as the technology of choice for the design of advanced integrated circuits. Furthermore, it is the backbone technology for the electronics of ICT resources such as servers, storage devices, and networking equipment in data centers.

Looking ahead, researchers are exploring new avenues for improving CMOS technology, such as the development of alternative materials, novel device architectures, and innovative manufacturing techniques. These efforts aim to address the challenges of scaling down feature sizes to the nanometer scale, while also meeting the increasing demands for performance, energy efficiency, and reliability in future generations of integrated circuits. Overall, the continued evolution of CMOS technology, coupled with ongoing research into new materials, device structures, and system architectures, holds great promise for the future of integrated circuit design and semiconductor technology.

6.1.2 Demand Calculation

When delving deeper into the power demand of CMOS circuits, it's essential to consider dynamic and static power components. Dynamic power is associated with the switching activity of transistors within the circuit. This component of power demand occurs when transistors switch states, leading to charge/discharge of capacitances and energy dissipation. On the other hand, static power, also known as leakage power, is the power used by the circuit even when it is not actively switching. This power component is influenced by factors such as subthreshold leakage, gate leakage, and junction leakage currents (Butts and Sohi 2000).

As technology nodes shrink (see Sect. 6.1.1), static power becomes a more significant portion of the overall power demand and energy consumption due to increased leakage currents in smaller transistors. Design techniques such as power gating, voltage scaling, and transistor sizing can be employed to mitigate both dynamic and static power demand in CMOS circuits. By carefully balancing these factors and optimizing the circuit design, engineers can achieve efficient and low-power implementations for a wide range of electronic devices and systems.

The authors in Butts and Sohi (2000) proposed a model for predicting the static power demand of CMOS circuits as:

$$P_{\text{CMOS}}^{\text{idle}} = V \cdot t \cdot k_{\text{design}} \cdot I_{\text{leak}} \tag{6.1}$$

where V is the supply voltage, t denotes the number of transistors, and k_{design} is a constant such that those three parameters are dependent on the design of the CMOS circuit, whereas I_{leak} is the normalized leakage current whose value is based on technology dependent parameters.

On the other hand, the dynamic power demanded by CMOS circuits can be calculated using the following equation (Chandrakasan and Brodersen 1995):

$$P_{\text{CMOS}}^{\text{dynamic}} = \alpha \cdot C \cdot V^2 \cdot f \tag{6.2}$$

where α is the activity factor that accounts for the percentage of time that each node in the circuit switches between logic states, C denotes the total capacitance of the circuit, which is determined by the size[1] and layout of the transistors and interconnects, whereas V and f are the operating voltage and frequency of the circuit respectively.

By plugging in the values for capacitance, voltage, frequency, and activity factor, one can calculate the dynamic power demand of a CMOS circuit. This calculation is crucial for designing energy-efficient circuits and optimizing power demand in electronic devices. In Sect. 6.3, we showcase the provision of power prediction models, derived from Eqs. (6.1) and (6.2) for the most power-demanding components of servers such as multi-core CPUs, RAMs, and hard disks, among others by considering the von Neumann architecture which is described next.

6.2 Von Neumann Architecture for Computing

6.2.1 A Brief History

The von Neumann architecture is named after John von Neumann, a Hungarian-American mathematician and physicist who played a crucial role in the development of early computing systems. In the mid-1940s, von Neumann was part of a team at the institute for advanced study in Princeton, New Jersey, that worked on the

[1] A smaller number of transistors typically demand less power due to reduced capacitance and shorter switching times.

electronic numerical integrator and computer (ENIAC) (Goldstine and Goldstine 1946; Hartree 1946), one of the earliest electronic general-purpose computers. Him, along with other prominent figures such as Alan Turing (Cooper and Van Leeuwen 2013), proposed a new architecture that separated the program instructions and data, which was a significant departure from earlier designs.

Prior to this, computers had separate memory for data and instructions, making programming more complex. The von Neumann architecture, as outlined in a paper by von Neumann in 1945, laid the foundation for the modern stored-program computer. This architecture's key idea was to store program instructions in the same memory as data, allowing for more flexible and efficient computing (Aspray 1990).

The first computer to fully embody the von Neumann architecture was the electronic discrete variable automatic computer (EDVAC) (Von Neumann 1993), which was completed in the early 1950s. From then on, this architecture became the standard model for computer design and has been fundamental to the development of computers we use today, which is clarified next.

6.2.2 Unveiling Essence

The architecture which is also known as model, is a theoretical framework for a computer design that serves as the basis for most modern computers. It consists of four main components:

- Central processing unit (CPU): This is the brain of the computer where all processing takes place. It includes the arithmetic logic unit (ALU) for mathematical operations and the control unit (CU) for managing instructions.
- Memory: In the von Neumann architecture, both data and instructions are stored in the same memory unit. This is known as the von Neumann bottleneck, where the CPU and memory share the same bus for communication.
- Input/output (I/O) devices: These devices allow users to interact with the computer, such as keyboards, monitors, printers, and external storage devices like USB drives.
- Bus: The bus acts as a communication pathway that allows data and instructions to flow between the CPU, memory, and I/O devices.

Embodying the von Neumann architecture, the process of launching a program on your computer involves the CPU retrieving program instructions from memory, executing them via the ALU, and subsequently storing the results back in memory or displaying them on output devices like monitors. This foundational architecture has significantly shaped the design of contemporary computers, optimizing data processing and storage efficiency. For a deeper exploration of computing history, readers can consult reference (Ceruzzi 2003).

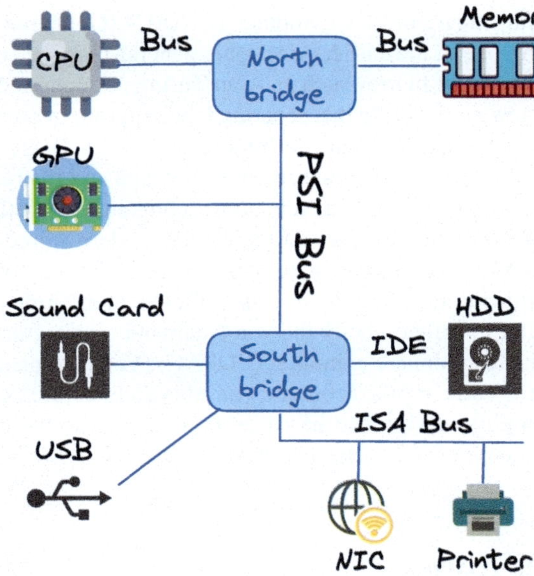

Fig. 6.2 The von Neumann architecture or model refers to a computer design that uses a single data bus to transfer data between the CPU, memory, and I/O devices. The Northbridge connects the CPU to high-speed devices, primarily the memory (e.g., RAM) and graphics card (e.g., GPU), and handles the communication between them to ensure smooth data flow. The Southbridge connects the slower peripheral devices like USB ports, hard drives (e.g., HDD), network interface card (NIC), and audio devices to the rest of the system, and handles communication between them thus facilitating their interaction with the rest of the computer

Figure 6.2 illustrates the essential design of the von Neumann architecture. It is important to highlight that the CPU interfaces with a system bus, segmented by North and Southbridges. Notably, the main memory (RAM) as well as the graphics card (e.g., GPU) link to the Northbridge through buses, emphasizing high-speed communication between CPU and those two components.

On the other hand, other peripheral devices such as USB, sound card, network interface card (NIC), and printer, which require slow communication are connected to the Southbridge. In essence, the Northbridge and Southbridge work together to manage data flow within the computer, with the Northbridge focusing on high-speed connections with critical components like the CPU, GPU and memory, while the Southbridge handles slower peripheral devices such as USB, NIC, audio card, and printer.

6.2.3 *Concisely Elaborating Core Elements of Modern Computing*

In this section, we provide a succinct overview of the key components of the von Neumann architecture as depicted in Sect. 6.2.2. Specifically, we introduce the processor (e.g., CPU), memory (e.g., RAM), storage (e.g., HDD and SSD), and the network interface card (e.g., NIC). Understanding these components will form the basis for comprehending the various power demanding models during both idle (e.g., static) and active (e.g., dynamic) states outlined in Sect. 6.3.

6.2.3.1 Processor

The evolution of multi-core processors has revolutionized computation, especially in parallel computing. In a multi-core processor, numerous cores reside on a single chip, enabling concurrent task processing. Each core boasts its dedicated on-chip cache for storing fetched information from memory. Additionally, cores on the same die share an off-chip cache, facilitating data sharing and coordination for diverse task segments. Inter-core communication within the same die and across different dies is made possible through system buses.

The voltage regulator plays a crucial role in adjusting the voltage for various cores. In scenarios where a single voltage regulator is employed, as depicted in Fig. 6.3, voltage adjustments impact all cores across different dies. Conversely, each die may feature its voltage regulator, allowing cores on different dies to operate at distinct voltage levels. We will see in Sect. 6.3 that the DVFS technology through the voltage regulator adjusts the voltage supplied to the processor cores dynamically to match the changing power requirements, thus the importance of such a regulator in modern multi-core processors.

6.2.3.2 Memory

Random access memory (RAM) technology has evolved significantly over the years to provide faster speeds, larger capacities, and improved efficiency. The earliest RAM technology was magnetic core memory, used in the 1950s and 1960s, which used tiny magnetized rings to store data (Papian 1953). Dynamic RAM (DRAM) was introduced in the 1970s, where data was stored in capacitors, requiring constant refreshing to maintain data integrity (Dennard 1984).

The synchronous DRAM (SDRAM) was proposed in the early 1990s as an improved version of DRAM which synchronized memory operations with the system clock, offering faster speeds and better performance (Jones et al. 1992). The double data rate (DDR) RAM technology was presented in the early 2000s, which doubled the data transfer rate of SDRAM and improved its efficiency and performance (Cosoroaba 1997).

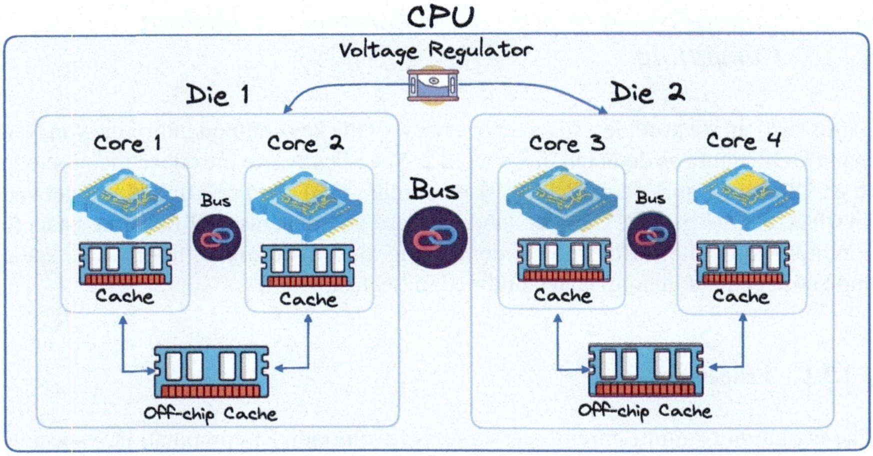

Fig. 6.3 A modern multi-core processor consists of (1) several cores where each core is an independent processing unit capable of executing instructions, (2) cache memory which is included in each core as well as in each die to store frequently accessed data and instructions, reducing the need to fetch them from slower main memory, (3) an interconnect fabric or bus which allows communication between different cores on the same processor chip or between cores of different dies, and (4) voltage regulator that plays a vital role in ensuring stable and efficient operation, where it is responsible for regulating the voltage supplied to the processor cores and other components on the chip

Subsequently DDR2, DDR3, DDR4, and DDR5 were proposed as subsequent generations of DDR SDRAM, which continued to increase data transfer speeds, reduce power demand, and enhance memory capacity. Due to these reasons, today DDR4 is the prevailing SDRAM technology in data centers, thus making it a staple in server environments, crucial for managing heavy workloads effectively.

In terms of architecture, a DDR4 memory module is segmented into bank groups, each housing one or more banks such that each bank is organized as an array consisting of rows and columns hosting multiple memory cells as well as having its own row as well as column decoders. Those decoders help to select the specific row and column of those memory cells within that corresponding bank during read and write operations.

This arrangement enables simultaneous access to various memory locations within a bank group, heightening memory access efficiency. Additionally, memory cells are tasked with storing data through electrical charges. Each cell comprises a capacitor for charge storage and a transistor for data manipulation.

As mentioned previously, conceptually the different memory cells function as an array of rows and columns, which can be accessed for data read/write operations using row and column address strobe signals (RAS and CAS), which are stored in their corresponding registers. RAS is triggered first to choose a row, followed by CAS for column selection within that designated row.

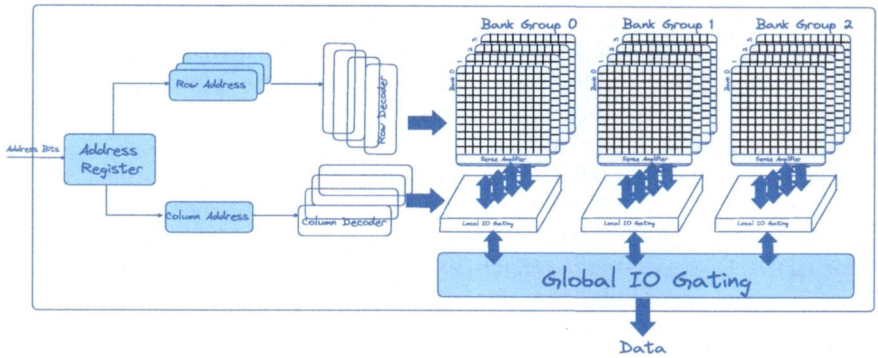

Fig. 6.4 Consider a DDR4 memory module containing 3 sets of bank groups (e.g., Bank Group 0 – 2), with each group housing 4 banks. Each bank is arranged as an array with rows and columns. Every bank is equipped with its column and row decoders, aiding in pinpointing the specific memory cell holding the data. Each bank group has dedicated row address registers, while the quantity of column address registers varies based on the memory module's density and design

Technically, each memory bank has its own row address register, whereas the number of column address registers depends on the architecture and density of the memory module. An address register is a component that stores the row address and column address strobe signals for accessing data stored in the memory cells. DDR4 memory modules use address registers to manage memory addressing and data retrieval efficiently.

When a memory read or write operation is initiated, the address registers help in selecting the correct row and column addresses to access the desired data stored in the memory cells. Figure 6.4 illustrates an example of a DDR4 memory module composed of 3 bank groups, where each group consists of 4 banks divided into rows and columns as explained above.

6.2.3.3 Storage

There are two distinct technologies employed in servers for non-volatile storage: The hard disk drive (HDD)—an older, less efficient technology, and the solid state drive (SSD)—a newer, more efficient option. The main components of an HDD commonly used in servers include platters, read/write heads, actuator arm, spindle motor, controller board, and interfaces or ports. Platters are the circular disks inside the HDD where data is magnetically stored.

Servers typically have multiple platters stacked on a spindle to increase storage capacity. The heads move rapidly over the spinning platters to read and write data. They are crucial for data retrieval and storage operations within the HDD. The actuator arm supports the read/write heads and positions them over the correct location on the platters. It moves the heads swiftly to access data on different parts of the disk.

The spindle motor spins the platters at high speeds, typically 7200 or 10000 RPM, enabling quick data access. The rotational speed affects the drive's performance and power demand. The controller board manages data transfer between the HDD and the server, processes commands, and controls various functions of the drive. It also contains firmware for drive operation.

The data interface of an HDD is the connection point for transferring data between the drive and the server's storage controller. Common interfaces include serial advanced technology attachment (SATA) and serial attached SCSI (SAS) for server-grade HDDs. Higher-end server HDDs, like the Western Digital Ultrastar DC HC520 series, may feature SAS interfaces for enterprise-level performance and reliability.

The illustration in Fig. 6.5 visually represents the various components of a hard disk drive (HDD) that we discussed earlier. It highlights key elements such as the spindle motor, read/write heads, and platters. In Sect. 6.3, we will explore how these mechanical components play a crucial role in the overall power demand of HDDs. Notably, this design leads to higher energy demands, not only when the drives are in active operation but also during idle periods.

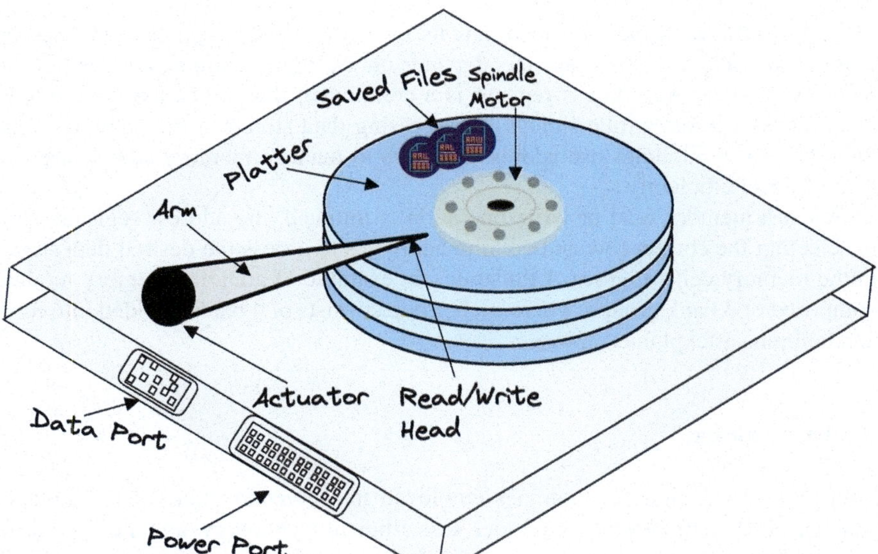

Fig. 6.5 An HDD consists of major components such as (1) platters which are circular disks coated with a magnetic layer where data is stored, (2) an actuator arm that holds the read/write heads and moves them across the platters, (3) read/write heads which are responsible for reading data from and writing data to the platters, and (4) spindle motor that spins the platters at high speeds for data access. Data is stored magnetically on the platters as 0s and 1s. The read/write heads move over the spinning platters to read or write data. The actuator arm positions the heads with precision. When data is read or written, the heads magnetize the particles on the platters. The spindle motor rotates the platters at a constant speed for access

SSDs work by using NAND flash memory chips (Aritome 2015), which are the primary storage components in an SSD where data is stored. NAND flash memory is non-volatile, meaning data is retained even when power is disconnected. When data is written to an SSD, it is stored in memory cells using electrical charges to represent binary data. To read data, the SSD retrieves the stored information by measuring the electrical charges in the memory cells.

This process allows for faster data access compared to traditional HDDs because there are no moving parts involved in reading or writing data. Servers may feature multiple NAND chips to increase storage capacity and read/write speeds. Some SSDs include a small amount of DRAM as a cache to temporarily store data being read or written. This cache helps improve overall performance by speeding up data access.

Like HDDs, SSDs connect to servers via interfaces such as SATA, peripheral component interconnect express (PCIe), or non-volatile memory express (NVMe). NVMe is particularly popular for high-performance server SSDs due to its low latency and high data transfer speeds. It is important to note that SSDs are more durable, energy-efficient, and provide faster performance than HDDs due to their design.

6.2.3.4 Network Interface Card

A network interface card (NIC) plays a crucial role in servers by enabling them to connect to a network and communicate with other devices. It serves as the interface between the server's internal components and the network infrastructure connecting other servers within a cluster of the same or different data center(s). For example, in a data center, a server NIC allows the server to send and receive data over Ethernet networks, enabling it to access resources, exchange information, and participate in network services such as file sharing, printing, and Internet connectivity.

There are primarily two types of NICs available: Ethernet and Wi-Fi. However, Ethernet NICs dominate the server landscape in data centers for several compelling reasons. First and foremost, data centers depend on Ethernet for its stability and low latency. Wired connections are far more reliable, facing significantly less interference than their wireless counterparts, making them ideal for mission-critical applications. Additionally, many data center applications demand high bandwidth and low latency. Wired Ethernet, especially at 10 G, 25 G, or even 100 G speeds, offers performance levels that Wi-Fi struggles to achieve.

Security is another crucial factor; wired connections are inherently more secure than wireless ones. In environments handling sensitive data, mitigating potential wireless vulnerabilities is paramount. Moreover, Wi-Fi NICs usually consume more power at idle compared to Ethernet NICs, and energy efficiency is a top priority for data centers to control operational costs.

Lastly, data centers are built with extensive cabling infrastructure to facilitate high-density deployments. Incorporating Wi-Fi capabilities would necessitate additional infrastructure, which may not be warranted given the existing setup. Overall, these factors collectively make Ethernet the preferred choice for NICs in data centers. Thus, throughout this chapter, any reference to a NIC for servers pertains specifically to Ethernet NICs.

A NIC in servers consists of several key components, including an Ethernet controller, ports, memory buffers, and drivers. The Ethernet controller manages the sending and receiving of data packets over the network. It controls the flow of data to and from the server over the network. Advanced features like offloading tasks such as TCP/IP checksum offload and segmentation offload can be integrated into the controller for improved network performance.

For example, the Mellanox ConnectX-4 Lx NIC incorporates an advanced controller with support for technologies like remote direct memory access (RDMA). Ports are physical connectors on the NIC that link the server to the network. For instance, a server NIC can have multiple ports to support different network speeds or connectivity options. These can include RJ45 ports for Ethernet connections, enhanced small form-factor pluggable (SFP+) ports for 10 Gigabit Ethernet, or even quad SFP (QSFP) ports for higher-speed connections.

Some server NICs, like the Intel X710, offer dual or quad ports to enhance networking capabilities and provide redundancy. The memory buffer on a NIC helps in temporarily storing incoming and outgoing data packets before processing. It aids in managing bursts of incoming traffic and ensures smooth data delivery. NICs with larger buffer sizes, such as the Broadcom NetXtreme E-Series, can handle higher volumes of network traffic without dropping packets or experiencing bottlenecks.

NIC drivers are essential software components that enable the NIC to communicate with the server's operating system (OS). They provide the necessary instructions for the OS to interact with the NIC hardware effectively. Updated drivers are crucial for maintaining optimal performance and compatibility. For instance, Intel offers specific drivers for their server NICs, like the Intel Ethernet Converged Network Adapter X710. These components collectively play a vital role in ensuring reliable, high-speed network connectivity for servers in various computing environments.

6.3 ICT Demand Breakdown: Servers Dominate Energy Usage in Data Centers

Data centers primarily serve as hubs for conducting various computations, such as training machine learning models, carrying out optimizations, or performing simulations, among others. Consequently, servers stand out as the most power-intensive ICT resources responsible for executing these calculations. Upon closer examination of server energy consumption, it becomes evident that processors account for approximately 60% of a server's energy usage, with the remaining 40% allocated to I/O operations (Basmadjian and De Meer 2012). These I/O activities encompass tasks like data transfer between memory and hard disks, as well as networking functions essential for communication purposes.

As outlined in Sect. 6.1.2, there are two distinct types of power demand occurring within CMOS circuitry: static and dynamic. In the case of dynamic power demand, it is essential to continuously monitor and transmit real-time data, such as current and voltage, to the predictive models underlying the system. A comprehensive explanation of the feedback loop can be located in Sect. 5.1, while Sect. 5.3 provides an illustrative high-level architectural overview and schema designed for optimizing energy consumption. Next, we will discuss methodologies used to devise power prediction models for servers when they are idle (e.g., inactive) or in dynamic states, such as during computing tasks.

6.3.1 Idle Demand of Servers

Essentially, the idle power demand of a server remains relatively stable (e.g., ±1 W) and is heavily influenced by its hardware configuration (e.g., number of processors, memory modules, fans). In recent years, various initiatives have been undertaken to decrease the idle power requirements of servers through the implementation of software solutions (e.g., C-states for processors, and sleep states for hard disks) alongside advancements in hardware design, architecture, and new technologies such as solid-state drives.

Following that, we will provide analytical power prediction models for each of the aforementioned significant components, taking into account various power management techniques such as dynamic voltage and frequency scaling (DVFS), sleep states, etc. These models will offer insights into the idle power demand behavior of individual components and aid in optimizing power efficiency across servers of data centers.

6.3.1.1 Processor

The introduction of multi-core processors[2] has brought about various power-saving mechanisms (e.g., C-states[3]) which significantly contribute to lowering the idle power demand of processors. A straightforward approach, which is explained next, to developing power prediction models for processors involves initially creating models without factoring in power-saving mechanisms, and subsequently incorporating them into the model as a reduction factor.

---

[2] The maximum number of cores in a modern CPU can vary, but as of now, some high-end CPUs can have up to 64 cores. However, this number is constantly evolving as technology advances.

[3] These states allow the CPU to reduce power demand by lowering its operating frequency or even putting parts of the processor to sleep when not actively processing tasks. This helps in conserving energy and improving the overall efficiency of the system.

Without Power-Saving Mechanisms

The power usage of a processor can be calculated using the formula based on Joule's and Ohm's laws as follows (Meade and Diffenderfer 2007):

$$P = V \cdot I, \tag{6.3}$$

where V and I denote the voltage (in Volt) and current (in Ampere) respectively. In the idle state of a multi-core processor, each of its individual cores is inactive. Equation (6.3) can be applied at the core level, assuming that each core equally contributes to the total idle power demand of the processor:

$$P_i = V_i \cdot I_i, \tag{6.4}$$

where P_i, I_i, and V_i represent the power, current, and voltage of the core i, respectively. The power demand of each core i is influenced by the number of transistors it contains, allowing for the estimation of idle power demand at the transistor level (see Eq. (6.1)). The greater the number of transistors in a core, the higher its idle power demand.

To calculate the power demand of transistors, Eq. (6.3) can be applied at the transistor level. For the jth transistor (in the order of millions) within the ith core, the power demand is determined by:

$$P_i^j = V_i^j \cdot I_i^j, \tag{6.5}$$

where I_i^j and V_i^j denote respectively the current and voltage of the jth transistor of the core i. It is important to note that the voltage V_i^j is typically equal to V_i of core i, which, in turn, matches the processor's voltage if a single voltage regulator is used for all cores, similar to the case presented in Fig. 6.3.

Figure 6.6 illustrates the current-voltage relationship of a MOSFET (Ortiz-Conde et al. 2002). Observing this figure reveals a linear correlation between current and voltage to the left of the red curve. Inspired by this observation, one can utilize the curve-fitting methodology to characterize the current leakage of a transistor by employing a second-order polynomial as:

$$I_i^j = \alpha_i \cdot (V_i^j)^2 - \beta_i \cdot V_i^j + \gamma_i, \tag{6.6}$$

where α_i, β_i, and γ_i are coefficients specific to core i that can be specified experimentally. Let t_i denote the total number of transistors (e.g., millions) of a core i, then its power demand can be estimated as:

$$P_i = \sum_{j=1}^{t_i} P_i^j = \sum_{j=1}^{t_i} V_i^j \cdot I_i^j = \sum_{j=1}^{t_i} \alpha_i \cdot (V_i^j)^3 - \beta_i \cdot (V_i^j)^2 + \gamma_i \cdot V_i^j \tag{6.7}$$

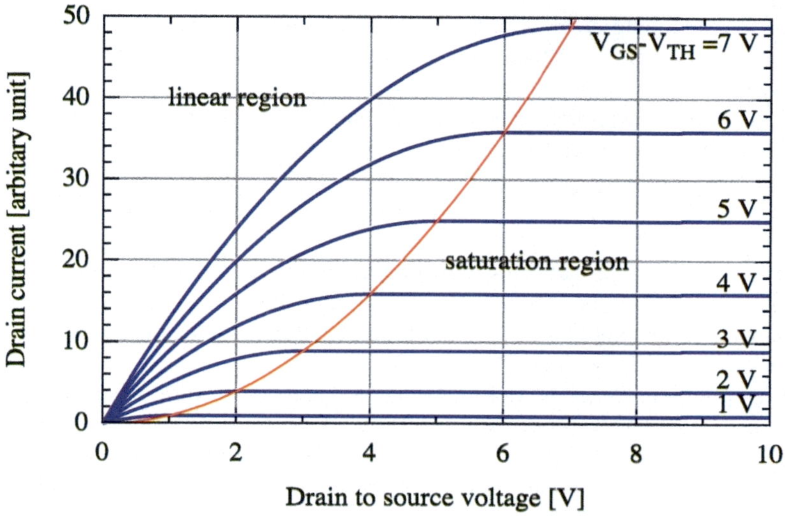

Attribution: Figure by CyrilB, licensed under CC BY-SA 3.0. Source: `https://commons.wikimedia.org/w/index.php?curid=15316422`

Fig. 6.6 The graph illustrates the current-voltage (IV) characteristics of a MOSFET. The X-axis represents the drain-to-source voltage (V_{DS}), while the Y-axis represents the current. The blue curves depict various gate-to-source voltage (V_{GS}) values. It is evident that different V_{GS} values result in two distinct regions: one exhibiting a linear relationship between voltage and current (referred to as the linear region), and the other showing a constant relationship known as the saturation region. Each blue curve has defining points for these regions, with the linear region to the left and the saturation region to the right. Connecting these points generates the red curve

where the power demand of the ith core is denoted as P_i and the other parameters are defined in Eq. (6.6).

With Power-Saving Mechanisms

Several power management mechanisms can be implemented when multi-core CPUs are in idle state. Clock and power gating techniques optimize power demand in this state. Clock gating selectively turns off clock signals to specific circuit blocks within the processor when not in use, reducing power demand. Power gating shuts down entire cores or functional units when unnecessary, further decreasing power demand. Dynamic voltage and frequency scaling (DVFS) adjusts processor voltage and frequency based on workload, saving power during idle periods.

Thread migration shifts active threads to specific cores, places other cores in low-power states, and redistributes workload efficiently to minimize power demand. Deeper C-states transition the processor to lower-power modes by selectively shutting down cores, reducing clock speeds, and lowering voltages. Managing temperature to prevent overheating enhances power efficiency by reducing reliance on power-intensive cooling systems.

Taking into account the factors mentioned above, the idle power demand of a core i having the power management mechanisms can be given as:

$$P_i^{\text{idle}} = (1 - \delta_i) \cdot P_i \tag{6.8}$$

where $\delta_i \in [0, 1]$ is the factor for reduction in the power demand P_i of core i, such that P_i is given in Eq. (6.7). It is worthwhile to note that since various power-saving mechanisms can be applied to different processor cores, then each core i may have its own power reduction factor δ_i.

Developing a universal model for δ_i that accommodates various core types with unique power-saving features is not just difficult but also unfeasible. Nevertheless, when considering the dynamic voltage and frequency scaling (DVFS) method, akin to Eq. (6.6), one can still employ a curve-fitting approach. A second-order polynomial equation can be suggested based on the core i voltage in the subsequent manner:

$$\delta_i = \alpha_i^f \cdot V_i^2 + \beta_i^f \cdot V_i + \gamma_i^f \tag{6.9}$$

where α_i^f, β_i^f, and $\gamma_i^f \in \mathbb{R}$ are coefficients specific to core i that can be specified experimentally, and V_i denotes the corresponding voltage of the core i. It is important to highlight that the coefficients α, β, and γ may vary across different frequency ranges. Therefore, a superscript f is appended to these coefficients to capture this variability.

As demonstrated in Basmadjian et al. (2012), frequencies within a specific range exhibit nearly identical reduction factors of δ_i. The precise number of these ranges is contingent upon the processor and its technology, which may require experimental specification. For example, in Basmadjian et al. (2012), the authors segmented into four frequency ranges due to the observed behavior of the processor being investigated.

Note that in case the adopted power-saving mechanism turns off completely the corresponding core i (e.g., $V_i = 0$), then it is possible to configure γ_i a value of 1 so that by replacing those values into Eq. (6.9), we obtain a value of 1 for δ_i, thus indicating zero power demand of that core i (see Eq. (6.8)).

Given a multi-core processor composed of n cores with specific energy-saving mechanisms, then its idle power demand is given by:

$$P_{\text{CPU}}^{\text{idle}} = \sum_{i=1}^{n} P_i^{\text{idle}} = \sum_{i=1}^{n}(1 - \delta_i) \cdot P_i = \sum_{i=1}^{n}(1 - \delta_i) \cdot (\sum_{j=1}^{t_i} \alpha_i \cdot (V_i^j)^3 - \beta_i \cdot (V_i^j)^2 + \gamma_i \cdot V_i^j) \tag{6.10}$$

where P_i^{idle} is the idle power demand of the core i by considering power-saving mechanisms and is given in Eq. (6.8).

It can be noticed from Eq. (6.10) that the only dynamic parameter that is required for the prediction is the voltage of each core i, whereas the other three parameters of α_i, β_i, and γ_i are static and can be derived and obtained through experiments, while δ_i can also be derived from the voltage and is given in Eq. (6.9).

6.3.1.2 Memory

Random access memory (RAM) demands power in its idle state required mostly to refresh specific ranks containing stored data. This section exclusively focuses on synchronous dynamic memories (SDRAM). Additionally, as mentioned in Sect. 6.2.3.2 currently double data rate 4 (DDR4) RAM technology is predominant in servers of data centers, as it offers increased performance, higher capacity, and improved energy efficiency compared to older generations like DDR, DDR2, and DDR3. Consequently, this section will provide an idle power prediction model for the DDR4 technology.

The idle power demand of a DDR4 memory module typically ranges from around 1 to 3 W, primarily based on its capacity and voltage. For instance, a standard DDR4 8 GB memory module has an idle power demand of around 1.5 W. In contrast, a higher capacity or faster memory module could have a slightly higher idle power demand. Thus, the demand can vary based on factors such as the specific manufacturer, module configuration (e.g., number of modules and their size), voltage and frequency settings, and particular DDR4 module specifications (e.g., refresh rate, capacitance), among others.

Delving into more details, the main power-demanding part of the memory during its idle state is due to the refreshing circuitry. Additionally, the refresh rate of DDR4 RAM modules is typically 8 K or 16 K refresh cycles every 64 ms. This ensures that data stored in memory cells is continuously refreshed to maintain the stability and integrity of the stored information.

It is important to note that the refresh rate for DDR4 RAM modules remains the same regardless of their size. Whether it's an 8 GB, 16 GB, or 32 GB DDR4 module, they all require the same refresh rate to maintain data integrity and stability. This can be mathematically given as:

$$P_{\text{RAM}}^{\text{idle}} = \frac{N_{\text{row}} \cdot R_{\text{act}} \cdot V^2 \cdot f}{T} \tag{6.11}$$

where N_{row} is the number of rows in the memory array and is directly related to the size or capacity of the memory module, R_{act} is the activation rate, V is the supply voltage (e.g., 1.2 V for DDR4), f denotes the frequency (e.g., between 2133 and 3200 MHz), and T is the refresh interval (e.g., 64 ms).

Setting the proper values to the aforementioned parameters, yields a power demand ranging from 1.2 to 2.95 W based on the specifics of the memory. For instance, let us consider an 8 GB (e.g., number of rows of 65536) 3.2 GHz DDR4

RAM with 1.2 V and an activation rate of 0.1 (e.g., 10%). This leads to a power demand of 1.473 W, which is necessary for refreshing the circuitry during idle time.

6.3.1.3 Storage

HDD's idle power demand can vary between 4 and 12 W, contingent on factors like capacity, operation mode (e.g., inactive, sleep, or standby), and technology. Conversely, SSDs generally exhibit lower power demand than HDDs due to their absence of moving mechanical components as mentioned in Sect. 6.2.3.3. The idle power usage of an SSD can range from 0.5 to 5 W (similar to a DDR4 memory module), depending on the capacity and the type of NAND flash memory utilized.

HDDs can have different power-saving modes in the idle state, such as inactive, sleep, and standby modes. These modes help reduce power demand when the drive is not actively being used:

- Inactive mode: In this mode, the HDD may spin down the disk platters and reduce power demand to a minimum while still being ready to quickly respond when needed. The drive remains connected and ready to resume normal operation rapidly.
- Sleep mode: This mode further reduces power demand by putting the drive into a low-power state where most components are powered off. The drive may take a bit longer to wake up from sleep compared to inactive mode.
- Standby mode: The drive is in a deep sleep state, demanding very little power, and may take longer to wake up compared to sleep mode. In standby mode, more mechanical components are powered down to save additional power and energy.

These power-saving modes in HDDs are designed to balance energy efficiency with responsiveness, ensuring that the drive can quickly return to an active state when needed while conserving power during periods of inactivity. Based on the aforementioned facts, the idle power demand of an HDD can be mathematically modelled as:

$$P_{\text{HDD}}^{\text{idle}} = P_{\text{HDD}}^{\text{base}} + \text{N} \cdot k_{\text{HDD}}^{\text{mode}} \tag{6.12}$$

such that $P_{\text{HDD}}^{\text{base}}$ is the base idle power demand (in W) of the HDD and N is the capacity (in GB), whereas mode indicates one of the three states of inactive, sleep or standby such that $k_{\text{HDD}}^{\text{inactive}}$, $k_{\text{HDD}}^{\text{sleep}}$, and $k_{\text{HDD}}^{\text{standby}}$ denote respectively the additional power demand per unit of storage capacity in the inactive (W/GB), sleep (W/GB), and standby (W/GB) modes.

For instance, assuming an HDD of 500 GB having an idle demand for inactive, sleep and standby modes of 0.01, 0.005, and 0.002 W/GB respectively as well as a base idle demand of 3 W, this leads to a total idle demand of 8 W, 5.5 W, and 4 W respectively when in inactive, sleep or standby modes.

SSDs do not have the same power states such as inactive, sleep and standby modes as traditional HDDs. This is because, unlike HDDs, SSDs do not have mechanical parts like spinning disks, so they don't have those modes that require varying levels of power demand based on disk activity. Consequently, SSDs demand relatively less power compared to HDDs, especially when idle, as they use flash memory to store data. This makes them more energy-efficient and faster in terms of data access. The idle power demand of SSDs can be mathematically presented as:

$$P_{\text{SSD}}^{\text{idle}} = P_{\text{SSD}}^{\text{base}} + N \cdot k_{\text{SSD}} \tag{6.13}$$

such that $P_{\text{SSD}}^{\text{base}}$ is the base idle power demand of the SSD (W) and N is the capacity (in GB), whereas k_{SSD} denotes the additional power demand per unit of storage capacity (W/GB). Given a 1TB SSD with a base power demand of 1 W and k_{SSD} of 0.001 W/GB, then the idle demand is about 2 W. Note that the exact values of the aforementioned parameters might be different based on the manufacturer and its specifications.

6.3.1.4 Network Interface Card

The idle power demand of an Ethernet NIC in servers can vary depending on the specific model and manufacturer. For example, a typical Gigabit Ethernet NIC might have an idle power demand of around 0.5–1 W, whereas a 10 Gigabit Ethernet NIC could demand around 1–2 W at idle.

Several factors can influence the idle power demand of a NIC in servers of data centers:

- Bandwidth: Higher-speed NICs (e.g., 10 G, 25 G) often have a higher idle power draw due to more complex circuitry and power requirements.
- Number of ports: Each additional port typically adds to the overall power demand of the NIC. For example, a dual-port 10G NIC might idle at 3–4 W, while a single-port version might be around 1–2 W. The power required can be roughly additive, although efficiency may vary based on design.
- Power management features: NICs supporting energy-efficient Ethernet can enter low-power modes during idle periods, significantly reducing power demand. For example, an EEE-enabled 10 G NIC may drop its idle power demand from 1.5 W to under 0.5 W when inactive.
- Manufacturing process: Newer, smaller fabrication technologies (e.g., 7 nm vs. 14 nm) generally lead to better energy efficiency. A NIC manufactured on a 7 nm process will often consume less power at idle than an older 28 nm counterpart.
- Driver optimization: Well-optimized drivers can facilitate better power management, allowing NICs to enter lower power states more effectively when idle. For instance, a NIC with poorly optimized software might consume 1W at idle, while a well-optimized one could reduce that to 0.3 W.
- Network activity: The amount of network traffic and the NIC's workload can affect its power demand, even during idle times. For instance, continuous polling for data

or maintaining connections can prevent the NIC from entering low-power states, leading to higher idle demand.

- Power management settings: The configuration of power management settings in the server's operating system can influence how the NIC behaves during idle periods.
- Environmental conditions: Factors like temperature and airflow within the server can also impact the power demand of the NIC.

Deriving an analytical model that incorporates all the aforementioned factors is quite complex. One effective approach to model the idle power demand of a NIC is to start with a linear equation that focuses on the most significant factors, excluding energy efficiency. Then, the total idle demand can be adjusted by applying a reduction factor, such as:

$$P_{\text{NIC}}^{\text{idle}} = (1 - \alpha) \cdot (P_{\text{NIC}}^{\text{base}} + \sum_{\text{t}} P_{\text{Port}}^{\text{idle}} + P_{\text{Comp}}^{\text{idle}}) \tag{6.14}$$

where $P_{\text{NIC}}^{\text{base}}$ represents the base power demand of the NIC when no additional factors are considered. This is the minimum power required by the NIC when powered on but not actively processing data. It can vary based on the NIC's design and chipset, typically ranging from 1 to 3 W for standard cards. $P_{\text{NIC}}^{\text{port}}$ denotes the power demanded by each port at idle, whereas t is the number of such ports. $P_{\text{Comp}}^{\text{idle}}$ gives the idle power demand of other features like offloading capabilities, LED indicators, or advanced circuitry.

Let us consider three cards with the specifications of dual-port 1 G, quad-port 10G, and single-port 25 G NICs respectively. Furthermore, let us assume that the base power demand for all three cards is $P_{\text{NIC}}^{\text{base}} = 1.5\,\text{W}$, where the idle power demand of the port is 0.5, 0.8, and 1 W respectively. Note that the idle power demand per port increases with the bandwidth and speed.

Finally, let us consider that each of the three cards has an idle demand for the components $P_{\text{Comp}}^{\text{idle}} = 0.3\,\text{W}$ and that the energy reduction ratio is 20%. Plugging those values into Eq. (6.14) leads to an idle power demand of 2.24, 4, and 2.24 W for the three considered Ethernet NICs respectively. Please bare in mind that the values adopted in this example are for demonstration purposes.

6.3.1.5 Motherboard

A motherboard in a server is the main circuit board that acts as the central nervous system of the server. It provides the physical foundation for connecting and integrating various components such as the CPU, RAM, storage devices, network interface cards, and other peripherals. For instance, in a server, the motherboard facilitates communication between the CPU and other components, manages data flow between devices, and houses essential connectors like PCIe slots for expansion cards.

It also includes features specific to servers, such as multiple CPU sockets, high-speed memory support, and advanced networking capabilities. Essentially, the motherboard in a server serves as a critical platform that enables all the hardware components to work together efficiently and reliably to deliver the server's computing power and services.

The idle power demand of the server (e.g., motherboard together with its constituent components) is given by:

$$P_{\text{Server}}^{\text{idle}} = \sum_{i=1}^{n_{\text{CPU}}} P_{\text{CPU}}^{\text{idle}} + \sum_{i=1}^{n_{\text{RAM}}} P_{\text{RAM}}^{\text{idle}} + \sum_{i=1}^{n_{\text{HDD}}} P_{\text{HDD}}^{\text{idle}} + \sum_{i=1}^{n_{\text{SSD}}} P_{\text{SSD}}^{\text{idle}} + \sum_{i=1}^{n_{\text{NIC}}} P_{\text{NIC}}^{\text{idle}} + c_i \tag{6.15}$$

where $P_{\text{CPU}}^{\text{idle}}$, $P_{\text{RAM}}^{\text{idle}}$, $P_{\text{HDD}}^{\text{idle}}$, $P_{\text{SSD}}^{\text{idle}}$, and $P_{\text{NIC}}^{\text{idle}}$ are given in Eqs. (6.10), (6.11), (6.12), (6.13), and (6.14) respectively, and c_i is a constant representing the additional power demand required by the motherboard, which can be derived statistically based on the server type.

6.3.2 *Dynamic Demand of Servers*

In contrast to the consistent idle power of servers, dynamic demand varies considerably depending on the active applications (e.g., jobs, tasks) and their respective workloads. Efforts have been made to reduce the dynamic power demand of servers through power-saving mechanisms like dynamic voltage and frequency scaling (DVFS) in conjunction with hardware architecture optimizations (e.g., ARM-based processors). Subsequently, the methodology for deriving prediction models for the dynamic power demand of key components within a server, including the processor, memory, and storage, is outlined next.

6.3.2.1 Processor

The power demand of the ith core of a CMOS-based multi-core processor due to performing certain computations is given by:

$$P_i^{\text{dynamic}} = \alpha \cdot C \cdot V_i^2 \cdot f_i \tag{6.16}$$

where all those parameters are defined in Eq. (6.2). Given a multi-core processor of n cores, then its dynamic power demand is given by:

$$P_{\text{CPU}}^{\text{unoptimized}} = \sum_{i=1}^{n} P_i^{\text{dynamic}} \tag{6.17}$$

where P_i^{dynamic} is the dynamic power demand of the ith core given in Eq. (6.16).

Regarding the dynamic power model presented in Eq. (6.17), a couple of observations can be made. Firstly, the equation integrates power-saving features such as dynamic voltage and frequency scaling (DVFS), where the dynamic power model for each core i, as indicated in Eq. (6.16), relies on voltage and frequency factors.

Consequently, increasing voltage and frequency results in higher dynamic power demand, while reducing these aspects leads to a decrease in power demand. To maximize energy savings, the power-saving mechanism adjusts the frequency and voltage to the lowest feasible values, thereby minimizing the power demand of the particular core i.

Secondly, it has been identified that the model in Eq. (6.17) tends to overestimate the dynamic power demand by approximately 20–30%. Research in Basmadjian and De Meer (2012) has demonstrated that the total dynamic power demand of a multi-core processor is not simply the sum of the power demands of its individual cores. This discrepancy arises due to resource sharing, such as the cache in certain multi-core processors, which diminishes power requirements.

By sharing cache, cores can access data more efficiently without needing to fetch it from external memory, which demands more power. This reduces the overall power demand of the processor because data is readily available within the shared cache, leading to improved performance and energy efficiency. For instance, in a multi-core processor with shared cache, if one core has already loaded data into the cache, other cores can access that data without fetching it again. This minimizes the need to access slower off-chip memory, thereby saving power and improving overall system efficiency.

Thanks to the aforementioned power-saving possibilities, a multi-core processor consistently demands less power compared to the Eq. (6.17). To address this overestimation, a power reduction factor δ (similar to the Eq. (6.8)) can be incorporated into Eq. (6.17) as follows:

$$P_{\text{CPU}}^{\text{dynamic}} = \delta \cdot P_{\text{CPU}}^{\text{unoptimized}} \tag{6.18}$$

where $P_{\text{CPU}}^{\text{unoptimized}}$ is the unoptimized dynamic power demand given by the Eq. (6.17) and $\delta \in [0, 1]$ is the power reduction factor. It's worth noting that a comprehensive analysis of multi-core processors goes beyond the scope of this chapter. Readers seeking a detailed understanding of deriving values for δ can refer to Basmadjian and De Meer (2012) for further insights.

6.3.2.2 Memory

In Sect. 6.3.1.2, it was highlighted that memory modules when idle primarily require power during self-refresh cycles. During these cycles, the memory controller might read and refresh multiple rows of memory cells to maintain data integrity. Although the refresh circuitry does involve read operations, the power demand related to these reads is minimal compared to the dynamic power needed especially when accessing large amounts of data or performing complex memory operations.

Several factors impact the dynamic power demand of DDR4 memory modules such as:

- Operating voltage: Higher operating voltage leads to increased power demand.
- Clock frequency: Higher clock frequencies result in more frequent data transfers and higher power demand.
- Memory capacity: Greater memory capacities require more power due to increased capacitance and the need to store and retrieve larger amounts of data.
- Data bus width: Different widths can influence power usage during read and write operations.
- Process technology: Advancements in process technology can influence power efficiency.

Drawing from the power requirements outlined in Eq. (6.2) for CMOS circuits and the relevant factors, the dynamic power demand of a DDR4 memory module can be comprehensively represented by the model:

$$P_{\text{RAM}}^{\text{dynamic}} = \alpha \cdot C \cdot V^2 \cdot f \tag{6.19}$$

In this model, each parameter aligns with those defined in Eq. (6.2), effectively integrating voltage and frequency components. As a result, this equation meticulously considers both voltage and frequency, encapsulating the first two essential factors mentioned earlier. When examining memory capacity and data bus width's impact on modeling capacitance of Eq. (6.19), considering a linear model proves beneficial. The following equation simplifies this relationship as follows:

$$C = \beta_0 + \beta_1 \cdot N + \beta_2 \cdot f + \beta_3 \cdot D_{\text{Width}} \tag{6.20}$$

where β_0, β_1, β_2, and β_3 are the coefficients that should be determined through statistical analysis, such as linear regression or through machine learning. Additionally, D_{Width} refers to the data bus width in bits, N signifies the memory module capacity in GB, and f is the frequency in GHz.

Let's examine 6 different DDR4 memory modules from the same manufacturer: the first one features a 32-bit bus width, 2.4 GHz of frequency, and a capacity of 4 GB, the second has 64-bit bus width, 3.2 GHz of frequency, and a capacity of 8 GB, the third possesses 128-bit bus width, 3.6 GHz of frequency, and a capacity of 16 GB, the fourth has a 256-bit bus width, 4 GHz of frequency and a capacity of 32 GB, the fifth features 128-bit bus width, 3.8 GHz of frequency, and a capacity of 64 GB, and the sixth boasts 256-bit bus width, frequency of 4.2 GHz, and a capacity of 128 GB.

For illustrative purposes, assume $\beta_0 = 15\,\text{pF}$, $\beta_1 = 0.003\,\text{pF/GB}$, $\beta_2 = 0.001\,\text{pF/MHz}$ and $\beta_3 = 0.002\,\text{pF/bit}$. It's essential to note that these values may vary depending on the manufacturing process and the manufacturer. By substituting these values into Eq. (6.20), the capacitance values for the first to sixth memory modules are calculated as 27.688 pF, 33.032 pF, 37.464 pF, 42.896 pF, 39.832 pF, and 46.584 pF respectively. Given an activity factor ($\alpha = 0.05$), and all the six memory modules operating at a voltage of $V = 1.2$, inserting those values together with the values of the capacitance calculated above for the six memory modules into the Eq. (6.19) yields to a dynamic power demand of 4.78 W, 7.6 W, 9.7 W, 12.3 W, 10.89 W, and 14.1 W respectively.

6.3.2.3 Storage

In Sect. 6.3.1.3, we observed that both HDDs and SSDs exhibit varying idle power demands. When comparing the power demand during data read/write tasks, HDDs usually require more power than SSDs due to their mechanical components. Conversely, SSDs generally have lower power requirements in both active and idle states as they lack moving parts and rely on flash memory for data storage. For example, a standard HDD might consume 6–10 W during active use, while an SSD may only need around 2–4 W for similar operations. It's important to note that these power values may differ depending on the specific make and model of the drives.

In general, writing data to an HDD demands more power compared to reading data from it. When writing data, the HDD's mechanical components, such as the spinning platters and moving read/write heads, are actively engaged, requiring more power to perform these operations. On the other hand, reading data involves fewer mechanical movements and typically demands less power in comparison. For example, when you copy large files onto an HDD, the power demand will be higher due to the constant movement of the read/write heads and spinning platters. However, when reading those same files back from the HDD, the power demand will be relatively lower.

A typical read operation on an HDD might demand around 6–7 W of power, while a write operation could consume around 8–9 W. These values can vary depending on factors such as the make, age and model of the HDD, as well as the rotational speed, efficiency, and workload. It is important to note that the capacity of the disk can influence the power demand of read and write operations on HDDs.

Generally, larger capacity drives may demand slightly more power due to the increased number of platters and heads required to store data. However, this difference in power demand is typically not significant for individual read and write operations. A straightforward analytical model for estimating the dynamic power demand of read and write operations of an HDD is given as:

$$P_{\text{HDD}}^{\text{dynamic}} = P_{\text{HDD}}^{\text{base}} + P_{\text{HDD}}^{\text{rpm}} + P_{\text{HDD}}^{\text{seek}} + P_{\text{HDD}}^{\text{platter}} \tag{6.21}$$

where $P_{\text{HDD}}^{\text{base}}$ is the base power demand of the HDD, $P_{\text{HDD}}^{\text{rpm}}$ is the power demand due to the rotation or spinning of the platters which is dependent on RPM, $P_{\text{HDD}}^{\text{seek}}$ is the demand required to move the head to the correct position, and $P_{\text{HDD}}^{\text{platters}}$ denotes the power needed to perform the writing of information on the platters. Note that for read operations both $P_{\text{HDD}}^{\text{seek}}$ and $P_{\text{HDD}}^{\text{platters}}$ have a value of zero.

When looking at the dynamic power demand of SSDs, even though those have no mechanical parts like HDDs, the write operations are more power-intensive than read operations due to the underlying technology and processes involved. More precisely, writing data to an SSD involves altering the charge state of NAND flash memory cells.

This process requires more power demand because higher voltages are needed to change the state of the cells compared to simply reading the charge state, which is relatively passive. Additionally, in NAND flash memory, data cannot be directly overwritten, as the existing data in a block must be first erased. This erasing process requires additional power, which involves resetting all cells in a block before new data can be written.

In addition to the aforementioned two reasons, error correction and garbage collection can also increase the power demand of write operations in SSDs. On the one hand, writing operations often necessitate more complex error correction processes. SSDs use error-correcting codes (ECC) to maintain data integrity, adding a computational load that increases power demand.

On the other hand, when writing, SSDs may also engage in garbage collection, where they clean up and reorganize data. This can involve moving data around, further increasing power demand during write cycles. So, while SSDs are faster and more efficient than HDDs in many ways, the nature of writing to flash memory inherently requires more power demand and energy consumption than reading from it.

The capacity of an SSD can influence its dynamic power demand both for read and write operations through various factors including parallel processing capabilities, increased management overhead, and buffer sizes. While larger capacities often lead to higher power demands, they can also provide efficiencies that help optimize overall energy use. Balancing these aspects is key in SSD design for both performance and power efficiency. Taking this fact into account, the dynamic power demand of read and write operations of an SSD can be modeled as:

$$P_{\mathrm{SSD}}^{\mathrm{dynamic}} = P_{\mathrm{NAND}}(N) + P_{\mathrm{controller}}(N) + P_{\mathrm{buffer}}(N) + P_{\mathrm{IO}} \tag{6.22}$$

where $P_{\mathrm{NAND}}(N)$ represents the power demand associated with read and write operations on NAND flash memory. This demand is directly proportional to both the number of NAND flash chips (e.g., T_{NAND}) and the storage capacity (e.g. N) of the disk. Meanwhile, $P_{\mathrm{controller}}(N)$ indicates the power required by the controller, which varies according to the disk's capacity N. The term $P_{\mathrm{buffer}}(N)$ accounts for the power demand of the memory buffer and can fluctuate with capacity. Finally, P_{IO} refers to the power utilized for input and output operations. Expanding these components in Eq. (6.22) results in:

$$P_{\mathrm{SSD}}^{\mathrm{dynamic}} = T_{\mathrm{NAND}} \cdot \alpha \cdot \frac{N}{N_{\mathrm{base}}} + (P_{\mathrm{controller}}^{\mathrm{base}} + \beta \cdot N) + \gamma \cdot N + P_{\mathrm{IO}} \tag{6.23}$$

where T_{NAND} is the number of NAND flash chips, α denotes the power coefficient per chip for read and write operations,[4] N_{base} is the reference capacity (e.g., 1 TB).

In short, the modeling of $P_{\text{NAND}}(N)$ assumes that power demand scales with the number of NAND chips associated with the capacity. In the second factor related to the demand of the controller, $P^{\text{base}}_{\text{controller}}$ denotes the base power of the controller, whereas β represents the increase in power per capacity (e.g., GB or TB). The third factor concerning the demand for buffer can be approximated by considering a linear relationship with respect to the capacity of the SSD. Values for the parameters of Eq. (6.23) can be chosen through experimental analysis.

6.3.2.4 Network Interface Card

The dynamic power demand of a NIC is shaped by the number of active ports operating at full capacity for packet transfer and their individual traffic rates. It's crucial to understand that when multiple ports are utilized, their power demand may not increase linearly due to shared components, much like we discussed earlier regarding Eq. (6.17). Thus, taking those facts into account leads to the following model:

$$P^{\text{dynamic}}_{\text{NIC}} = P^{\text{base}}_{\text{NIC}} + t \cdot (P^{\text{share}}_{\text{Port}} + P^{\text{overhead}}_{\text{Port}} + B(\alpha \cdot \rho + \beta)) \tag{6.24}$$

where $P^{\text{base}}_{\text{NIC}}$, $P^{\text{share}}_{\text{Port}}$, and $P^{\text{overhead}}_{\text{Port}}$ denote respectively the base demand of the NIC, the demand due to sharing resources, and demand due to the overhead of keeping the connection, whereas t is the number of the corresponding ports, $\rho \in [0, 1]$ represents the utilization rate, B denotes the maximum capacity of the port (e.g., 250 Mbps) such that α and β are the respective coefficients related to the corresponding traffic load. The coefficients of Eq. (6.24) can be obtained through experimental analysis.

Let's examine the same configuration outlined in Sect. 6.3.1.4 for three NICs: (1) two ports, each with a data rate of 1 Gbps, (2) four ports, each at 10 Gbps, and (3) one port with a data rate of 25 Gbps. We'll assume that each port operates at a 50% traffic load, with a base power demand of 2 W, and additional demands of 0.5 W for sharing and 0.1 W for overhead. Additionally, let's set α at 0.3 W/Gbps and β at 0.05 W/Gbps. Plugging these values into Eq. (6.24) results in dynamic power demands of 3.1 W, 11.9 W, and 7.1 W, respectively. Please note that these numerical values are intended solely for illustrative purposes.

---

[4] This can be higher for write operations due to reasons explained above such as programming cells, erase before write, error correction and garbage collection.

6.3.2.5 Motherboard

The dynamic power demand of the server (e.g., motherboard together with its constituent components) is given by:

$$P_{\text{Server}}^{\text{dynamic}} = \sum_{i=1}^{n_{\text{CPU}}} P_{\text{CPU}}^{\text{dynamic}} + \sum_{i=1}^{n_{\text{RAM}}} P_{\text{RAM}}^{\text{dynamic}} + \sum_{i=1}^{n_{\text{HDD}}} P_{\text{HDD}}^{\text{dynamic}} + \sum_{i=1}^{n_{\text{SSD}}} P_{\text{SSD}}^{\text{dynamic}} + \sum_{i=1}^{n_{\text{NIC}}} P_{\text{NIC}}^{\text{dynamic}} + c_d \quad (6.25)$$

where $P_{\text{CPU}}^{\text{dynamic}}$, $P_{\text{RAM}}^{\text{dynamic}}$, $P_{\text{HDD}}^{\text{dynamic}}$, $P_{\text{SSD}}^{\text{dynamic}}$, and $P_{\text{NIC}}^{\text{dynamic}}$ are given in Eqs. (6.18), (6.19), (6.21), (6.22), and (6.24) respectively, and c_d is a constant representing the additional power demand required by the motherboard, which can be derived statistically based on the server type.

6.3.3 *Total Demand*

Given a server consisting of one or more multi-core processors, memory modules of type SDRAM DDR4, hard disks and solid state drives, as well as network interface cards, then the total power demand can be estimated as:

$$P_{\text{Server}} = P_{\text{Server}}^{\text{idle}} + P_{\text{Server}}^{\text{dynamic}} \quad (6.26)$$

where $P_{\text{Server}}^{\text{idle}}$ and $P_{\text{Server}}^{\text{dynamic}}$ are the idle and dynamic power demand of the corresponding server given in Eqs. (6.15) and (6.25) respectively.

6.4 Summary

The chapter first highlights the importance of complementary metal-oxide-semiconductor (CMOS) technology in applications like microprocessors and data center infrastructure. It discusses ongoing research into alternative materials and manufacturing techniques that aim to overcome challenges at the nanometer scale, enhancing performance, energy efficiency, and reliability in integrated circuit design. Thus, the first part (Sect. 6.1) traces CMOS's evolution, beginning in the late 1960s with Chih-Tang Sah and Frank Wanlass at Fairchild Semiconductor, who developed it to improve upon earlier semiconductor technologies.

By using both p-type and n-type transistors, CMOS reduced power demand and enhanced noise immunity, gaining prominence over NMOS and PMOS by the mid-1970s, thanks in part to advancements by RCA and IBM. The section also covers significant developments in the 1980s and 1990s, particularly with deep submicron processes, leading to complex, energy-efficient circuits. It concludes with the presentation of two advanced analytical models for estimating power demand in CMOS circuits during idle and active states.

In the second part of this chapter (Sect. 6.2), we begin with a brief history explaining the emergence of the von Neumann architecture before delving into its design. This architecture employs a single data bus to facilitate data transfer between the CPU, memory, and I/O devices. The Northbridge connects the CPU to high-speed components, such as RAM and GPUs, ensuring seamless data flow.

Conversely, the Southbridge links slower peripherals, including USB ports, hard drives, network cards, and audio devices, enabling their interaction with the system. The section also provides an overview of the key components of the von Neumann architecture: the CPU, RAM, storage (HDD and SSD), and network interface card (NIC). Understanding these components is crucial for grasping the power-demanding models discussed in Sect. 6.3, which explore energy consumption during both idle (static) and active (dynamic) states.

The final section of the chapter examines power demand distribution in data centers, emphasizing that servers are the primary contributors to power demand and energy consumption. It presents methodologies for creating predictive analytical models for both idle and active power demands, focusing on key components of the von Neumann architecture, such as processors, memory, hard disk drives (HDDs), solid-state disks (SSDs), and network interface cards (NICs).

For the idle power demand, it notes a linear relationship between current and voltage in multi-core processors, allowing for a polynomial representation of power demand that factors in voltage and a power reduction factor. In terms of dynamic power demand, crucial variables include voltage, frequency, activity factor, and capacitance, all of which affect energy efficiency, with higher values typically leading to greater power needs.

Understanding these dynamics aids data center operators in managing energy use and optimizing server performance while balancing the trade-off between performance and energy efficiency. This necessitates advanced modeling techniques, such as Queuing Theory and Markov Chains, which will be explored in the next chapter.

References

Aritome S (2015) NAND flash memory technologies. Wiley

Arute F, Arya K, Babbush R, Bacon D, Bardin JC, Barends R, Biswas R, Boixo S, Brandao FG, Buell DA et al (2019) Quantum supremacy using a programmable superconducting processor. Nature 574(7779):505–510

Aspray W (1990) John von Neumann and the origins of modern computing. MIT Press

Basmadjian R, De Meer H (2012) Evaluating and modeling power consumption of multi-core processors. In: Proceedings of the 3rd international conference on future energy systems: where energy, computing and communication meet, pp 1–10

Basmadjian R, Niedermeier F, De Meer H (2012) Modelling and analysing the power consumption of idle servers. In: 2012 sustainable internet and ICT for sustainability (SustainIT), pp 1–9. IEEE

Bhuyan MH (2017) History and evolution of cmos technology and its application in semiconductor industry
Butts JA, Sohi GS (2000) A static power model for architects. In: Proceedings of the 33rd annual ACM/IEEE international symposium on Microarchitecture, pp 191–201
Ceruzzi PE (2003) A history of modern computing. MIT Press
Chandrakasan AP, Brodersen RW (1995) Minimizing power consumption in digital cmos circuits. Proc IEEE 83(4):498–523
Chih-Tang S (1988) Evolution of the mos transistor-from conception to vlsi. Proc IEEE 76(10):1280–1326
Cooper SB, Van Leeuwen J (2013) Alan turing: his work and impact. Elsevier
Cosoroaba AB (1997) Double data rate synchronous drams in high performance applications. In: WESCON/97 conference proceedings, pp 387–391. IEEE
Dennard RH (1984) Evolution of the mosfet dynamic ram-a personal view. IEEE Trans Electron Dev 31(11):1549–1555
Goldstine HH, Goldstine A (1946) The electronic numerical integrator and computer (eniac). In: The origins of digital computers: selected papers, pp 359–373. Springer
Gu RX, Elmasry MI (1996) Power dissipation analysis and optimization of deep submicron cmos digital circuits. IEEE J Solid-State Circuits 31(5):707–713
Hartree DR (1946) The eniac, an electronic computing machine. Nature 158(4015):500–506
Jones F, Prince B, Norwood R, Hartigan J, Vogley W, Hart C, Bondurant D (1992) Memory-a new era of fast dynamic rams (for video applications). IEEE Spectr 29(10):43–45
Jurczak M, Collaert N, Veloso A, Hoffmann T, Biesemans S (2009) Review of finfet technology. In: 2009 IEEE international SOI conference, pp 1–4. IEEE
Lin Y-S, Wu C-C, Chang C-S, Yang R-P, Chen W-M, Liaw J-J, Diaz CH (2002) Leakage scaling in deep submicron cmos for soc. IEEE Trans Electron Devices 49(6):1034–1041
Marković D, Mizrahi A, Querlioz D, Grollier J (2020) Physics for neuromorphic computing. Nat Rev Phys 2(9):499–510
Meade RL, Diffenderfer R (2007) Foundations of electronics: circuits and devices: conventional flow version. (No Title)
Ortiz-Conde A, Sánchez FG, Liou JJ, Cerdeira A, Estrada M, Yue Y (2002) A review of recent mosfet threshold voltage extraction methods. Microelectron Reliab 42(4–5):583–596
Papian WN (1953) The mit magnetic-core memory. In: Papers and discussions presented at the Dec 8–10, 1953, eastern joint AIEE-IRE computer conference: information processing systems—reliability and requirements, pp 37–42
Schaller RR (1997) Moore's law: past, present and future. IEEE Spectr 34(6):52–59
Shahidi GG (2007) Evolution of cmos technology at 32 nm and beyond. In: 2007 IEEE custom integrated circuits conference, pp 413–416. IEEE
Von Neumann J (1993) First draft of a report on the edvac. IEEE Ann Hist Comput 15(4):27–75

Chapter 7
Power and Performance Trade-Off Analysis

7.1 Dynamic Voltage and Frequency Scaling Technology

The DVFS technology is a critical mechanism integrated within the hardware architecture through on-chip regulators (Eyerman and Eeckhout 2011). This technology allows for the real-time adjustment of a processor's frequency and voltage in response to its immediate workload requirements. Modern multi-core processors are designed with multiple frequency settings, ranging from a minimum (f_n) of approximately 800 MHz to a maximum (f_0) of about 4.5 GHz. This variability ensures that the processor can operate efficiently across a spectrum of performance demands.

It's essential to highlight that hardware manufacturers not only provide these frequency options but also specify the corresponding voltage levels required for each frequency. Typically, this information is organized in an immutable table that maps frequencies to their respective voltages. Importantly, this table cannot be altered by software, including the operating system (OS), ensuring that the hardware maintains optimal performance levels without interference.

Modern multi-core processors often operate at much lower voltages for lower frequencies and higher voltages for higher frequencies (see Fig. 7 in Attia et al. 2017), with some values being in the range of 0.75 V at low frequencies, rising to around 1.35 V or slightly higher at maximum frequencies. Table 7.1 gives an example of this mapping table, where the values reflect a trend towards lower voltages at lower frequencies for energy efficiency, while higher frequencies correspond to increased voltages to maintain performance under load.

For the most accurate information, readers need to refer to the specific processor's datasheet or technical documentation. To fully leverage the capabilities provided by DVFS technology, specialized software known as a governor has been developed. This software operates at the kernel level of the OS and plays a crucial role in managing DVFS-enabled processors (Brodowski et al. 2013). The governor is responsible

R. Basmadjian, *Demand-Side Energy and Power Management in Data Centers*, Lecture Notes in Energy 1, https://doi.org/10.1007/978-3-031-85956-4_7

Table 7.1 An illustrative table showing the mapping of the frequency to the voltage defined by the manufacturer. Actual values depend on specific processor models and their design. For precise mappings, one would refer to the technical specifications provided by the manufacturer, such as Intel's or AMD's documentation for their respective processors, which detail the frequency-voltage relationship based on their architecture

Frequency [GHz]	Voltage [V]
0.8	0.75
1.2	0.85
1.6	1.0
2.0	1.10
2.4	1.2
2.6	1.25
2.8	1.30
3.0	1.35
3.2	1.4

for determining the most appropriate frequency for the processor to operate at, based on various predefined criteria.

These criteria can differ significantly depending on the type of governor implemented (see Sect. 7.1.1 for more details). For instance, the Linux operating system features five distinct governors: onDemand, Conservative, Performance, Powersave, and Userspace. Each of these governors employs unique strategies to optimize processor performance and energy efficiency (Pallipadi and Starikovskiy 2006), which we will explore in detail next.

7.1.1 Governors Explained

The Performance governor always sets the processor cores to the highest frequency, while the Powersave governor sets them to the lowest frequency (see the reason below). The Userspace governor allows for custom-defined switching rules based on the available frequency ranges of the processor defined between the minimum (f_n) and maximum (f_0) frequencies. The onDemand and Conservative are utilization-based governors, meaning they determine the switching rules based on the threshold of the utilization rate, which is typically monitored every 10 ms.[1]

The difference between these two governors lies in the scale of frequency jumps. When the average current utilization in the last 10 ms exceeds the set threshold (usually configured at 90%), the onDemand governor jumps to the highest frequency, while the Conservative governor jumps to the next higher frequency. Conversely,

[1] Is the default value. However, this can be configured to some other values.

when the current utilization rate is lower than the threshold, both governors decrease the frequency by one step until the lowest frequency is reached.

Figure 7.1 visually illustrates the selection process of the four governors discussed earlier, highlighting how they determine the most appropriate frequency for optimal performance. The color coding within the figure serves to represent varying power demand levels: red signifies high power demand, green indicates a focus on power efficiency, and orange represents a moderate power demand scenario. It's essential to understand that in this context, the color scheme also conveys performance levels: red reflects the highest performance capability, orange suggests a balanced approach with moderate performance, and green indicates a prioritization of energy efficiency, often at the expense of worse performance.

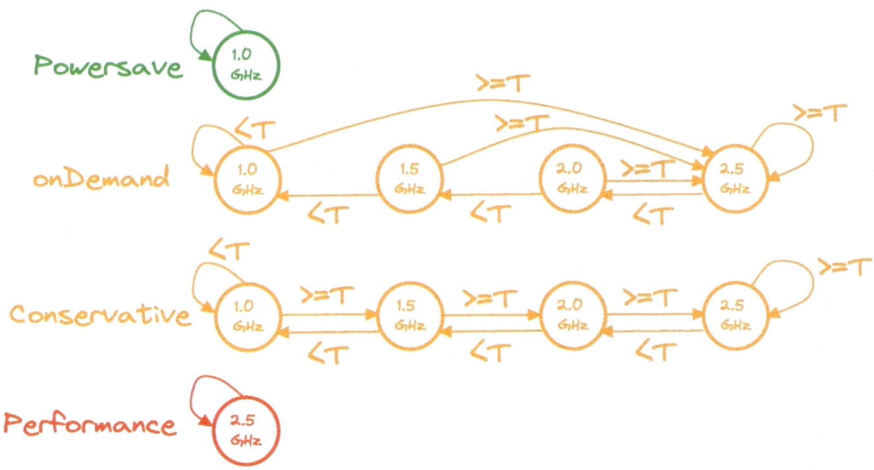

Fig. 7.1 This figure illustrates the operation of four basic governors in selecting the appropriate frequency. In this example, it is assumed that the processor supports four frequencies: 1.0, 1.5, 2.0, and 2.5 GHz. The colors represent power demand, with green indicating the lowest, orange denoting moderate demand, and red indicating the highest. The Powersave governor always sets and maintains the lowest frequency, while the Performance governor always sets and maintains the highest frequency. The onDemand and Conservative utilization-based governors differ in their selection of the next jumping frequency based on the configured threshold T. The onDemand governor always jumps to the highest frequency, while the Conservative governor jumps to the next higher frequency until the maximum is reached and can remain there if necessary. In both scenarios, the jump condition is determined by whether the utilization rate in the last 10 ms is equal to or higher than the specified threshold. This condition is represented in the figure as ">=T". Both governors gradually decrease the frequency until it reaches the minimum frequency, and if the utilization in the last 10 ms does not surpass the threshold T, the frequency can remain at the minimum level. This is indicated in the figure as "<T". The default values for the monitoring window (e.g., 10 ms) and the threshold T (e.g., 90%) can be customized according to specific requirements

7.1.2 Performance Versus Demand

DVFS technology aims to save power and minimize heat generation from the chip, particularly when the chip is operating at low to medium loads (Kuehn and Mashaly 2019). To comprehend the reasoning behind this concept, let's take a few moments to delve into it. The performance of running services is determined by the speed of the processor, which is measured by its frequency (in GHz).

In essence, the execution time T of a service is determined by the ratio of the workload S required to complete the service and the frequency f_i at which the processor is executing that particular service such that $f_i \in [f_n, f_0]$:

$$T = \frac{S}{f_i} \tag{7.1}$$

Equation (7.1) demonstrates that the optimal (e.g., shortest) execution time T for a given workload S is achieved when the processor operates at its maximum frequency f_0, resulting in the best performance. In terms of power demand P in CMOS circuits like processors, we saw in Chap. 6 that the dynamic power demand is proportional to the operating voltage V and frequency f as given in Eq. (6.2):

$$P = \alpha \cdot C \cdot V^2 \cdot f$$

where all the parameters are described in Sect. 6.1.2. Consequently, according to Eq. (6.2), as demand is directly related to frequency and voltage, increasing the values for those two parameters leads to higher power demand, and vice versa. By scaling the frequency and voltage up or down, the DVFS mechanism affects the power demand (increasing or decreasing it) and influences the performance of the running services (improving or degrading it).

7.1.3 Different Real-World Usages of the Technology

DVFS is a pivotal technology used in a variety of scenarios to optimize performance and energy consumption. For instance, in mobile devices, DVFS helps extend battery life by reducing the processor's voltage and frequency when the device is not under heavy load (Begum et al. 2015; Carroll and Heiser 2014; Kim et al. 2013). This means that tasks like checking notifications or browsing the web can be done efficiently without draining the battery quickly.

In computing systems, particularly in desktops and laptops, DVFS contributes to lowering cooling costs and minimizing noise (Halimi et al. 2013; Hanumaiah and Vrudhula 2012). By adjusting the voltage and frequency of the CPU based on current processing demands, the system can operate at lower power levels during idle or less intensive tasks. This not only reduces the heat generated, which in turn lessens the burden on cooling systems, but also leads to quieter operation since fans don't need to work as hard.

In data centers, DVFS plays a crucial role in managing power demand and energy consumption, particularly during periods of low workload utilization (Krzywda et al. 2018; Samual and Hussin 2018; Wang et al. 2017). Many servers can operate at lower frequencies and voltages when the demand for computing resources is low, significantly reducing the overall power demand. This is particularly beneficial as it can lead to substantial cost savings in electricity bills and help with the sustainability efforts of organizations.

However, to implement DVFS effectively, the underlying hardware of the equipment must be designed to support dynamic voltage and frequency adjustments. This includes not only the processors but also the power management architecture and software that governs these changes. As mentioned previously, the DVFS technology is advantageous for reducing power demand and energy consumption.

Nevertheless, it's important to note that it can also result in a decrease in performance for running services. Thus, the DVFS technology introduces a trade-off between power/energy efficiency and performance, necessitating careful consideration based on workload requirements. Next, we will present a methodology based on Markov chains and Queuing Theory that allow us to model and study this trade-off.

7.2 Queuing Theory Methodology

Queuing theory (Sundarapandian 2009) is a mathematical framework (e.g., methodology) used to model and analyze the behavior of queues, or waiting lines. It helps understand the dynamics of systems where entities (like customers, data packets, or tasks) arrive, wait for service, and then receive that service. Queuing theory is widely applicable in various fields including healthcare (Fomundam and Herrmann 2007), telecommunications (Giambene 2014), service industries (Haghighi and Mishev 2008), pharmacy (Nosek and Wilson 2001), manufacturing (Papadopolous et al. 1993), and quantum computing (Basmadjian and Paler 2023; Dobbs et al. 2021; Paler and Basmadjian 2019), among others.

7.2.1 Modelling Parameters

This methodology or mathematical framework consists of five key components (e.g., parameters), allowing for the modeling of different real-life problems, which are explained next:

1. Arrival process (A): The average arrival rate (λ) denotes the average number of customers (also known as jobs or tasks) arriving per time unit, say 12 customers per hour. For instance, in a bank, customers arrive randomly at the teller's window. The arrival process can often be modeled using a Poisson distribution, which assumes that the number of arrivals in a fixed interval of time (e.g., seconds, minutes, hours, etc.) follows a specific statistical pattern.
2. Service process (S): If a particular teller is busy, the next customer will have to wait, and the system's performance can be analyzed based on how long customers spend waiting versus how long they are being served. The average service rate (μ) indicates the average number of customers being serviced per time unit, say 5 customers per hour. Back to the example of the bank, each teller can serve customers at an average rate of 10 customers per hour (e.g., $\mu = 10$). The service times can be modeled using an exponential distribution, which is common in queuing models. For the example of teller processing with a rate of 10 customers per hour, then the service time of each customer would be on average 6 min.
3. Queue discipline (Q): It describes the order in which the customers or jobs inside the queue will be processed. The default queuing discipline is the well-known First-Come-First-Served (FCFS) meaning that the first customer to arrive is the first to be served. Other disciplines include Last-Come-First-Served (LCFS), priority queuing where certain customers get preferential treatment, or round-robin (RR). In a bank or grocery store, for instance, the queue discipline is FCFS.
4. Number of servers (m): It indicates the number of processing units available simultaneously for parallel processing, which can significantly reduce waiting times. For instance, a bank might have multiple tellers (servers), each capable of serving customers independently. If there are three tellers, the system can be modeled as an M/M/3-FCFS queue, where 'M' denotes a memoryless (Markovian) process for arrivals and service times, and '3' is the number of servers (e.g., tellers).
5. Queue capacity (K): It specifies the maximum number of customers or jobs in the queuing system including both in the queue and service. This limitation can affect customer satisfaction and the overall efficiency of the service process, because when the system is at its full capacity, then any arrival of new customers can not be serviced. In theory, there exist two possibilities: either the arriving job is lost, if the queue is full or sending nodes/processes are blocked. For instance, a call center might limit the number of callers on hold to 10 before sending additional callers to voicemail, creating a finite capacity queue (for example, M/M/1/K, where K is the maximum capacity).

By utilizing this set of five parameters, various queuing systems can be modeled, each reflecting a unique real-world scenario. The most basic model is the single-server queuing system with infinite capacity and an FCFS discipline, where both the arrival and service processes exhibit Markovian characteristics. This system is mathematically denoted as M/M/1-FCFS and is illustrated in Fig. 7.2.

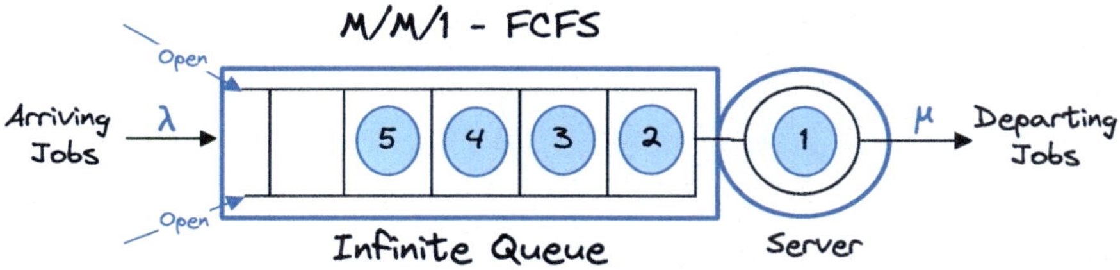

Fig. 7.2 The M/M/1-FCFS queuing system is designed to model scenarios featuring a single server with an infinite queue capacity. In this system, jobs (represented by blue circles) arrive at a rate of λ and depart at a service rate of μ, both following a Markovian process. The numbers within the blue circles indicate the sequence in which the jobs are arrived and serviced, adhering to the First-Come-First-Served (FCFS) discipline. This means that jobs are addressed in the order of their arrival, ensuring a fair and systematic processing of tasks

Notably, the graphical representation of infinite queue capacity is indicated by leaving the end of the queue open, as shown by the two blue arrows labeled "Open" in Fig. 7.2. Jobs enter the system at an arrival rate of λ and depart after being served at a service rate of μ. Please note that it is equally possible to set a bound to the queue of this system by specifying a given capacity, as we will see it next.

Figure 7.3 illustrates a queuing system with multiple servers and a finite queue capacity. In this setup, three servers can process three distinct jobs at the same time, while the underlying queue system can hold a maximum of eight jobs—three being actively processed and five waiting. It's important to note that the assignment of jobs to the different servers is arbitrary. Whenever one of the three servers becomes available, the next job in the queue is allocated to that free server. Like the previous queuing system, it is also possible to have an infinite-capacity multi-server queuing system.

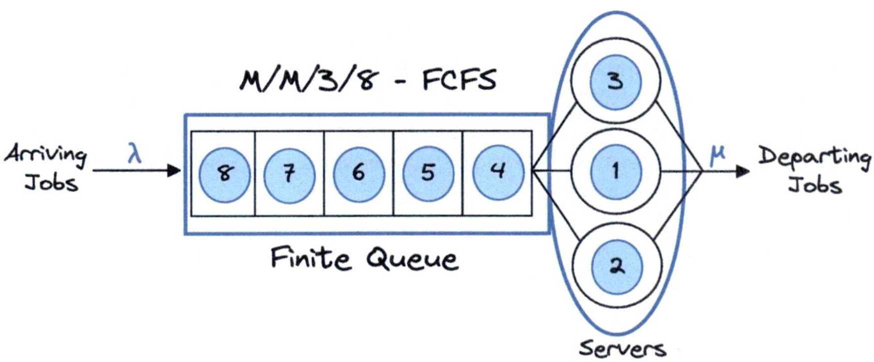

Fig. 7.3 The M/M/3/8-FCFS queuing system is designed to model scenarios featuring multiple servers (in this case it is 3) with a finite queue capacity (in this example 8 jobs). In this system, jobs (represented by blue circles) arrive at a rate of λ and depart at an overall service rate of μ, both following a Markovian process

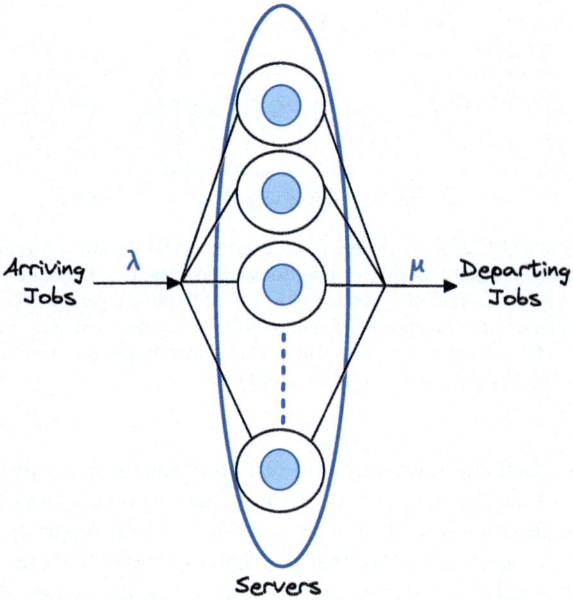

Fig. 7.4 Here's a graphical representation of an infinite server queuing system. In this model, incoming jobs are immediately assigned to free and available servers, thanks to the unlimited server capacity. This ensures a seamless process where every arriving job is promptly handled, eliminating the need for a temporary storage queue. This immediate assignment not only enhances efficiency but also improves the overall user experience, as there are no delays for jobs waiting to be processed. For example, in a tech support scenario, every customer call can be answered instantly, leading to higher satisfaction rates and streamlined operations

Another widely used queuing system model is illustrated in Fig. 7.4. In this scenario, the number of servers is limitless, allowing any incoming job to be immediately assigned to an available server. This instantaneous assignment effectively eliminates the need for a waiting queue, as there are no delays for jobs waiting to be processed. This model reflects an idealized environment where resources are abundant (e.g., data centers), ensuring that every job can be addressed without bottlenecks. By removing the queue entirely, this model showcases the advantages of having a flexible and responsive service infrastructure.

7.2.2 Performance Metrics in Queuing Theory

Once the queue is modeled using one of the aforementioned queuing system models, various performance metrics can be derived, such as:

1. Average number of customers in the system ($\bar{K}$): It indicates the mean number of jobs or customers in the queuing system at any time. For instance, if 12 customers arrive per hour and each spends an average of 30 min (e.g., 0.5 h) inside the bank, then $\bar{K}$ would be 6 (e.g., 12 × 0.5) customers on average. This is calculated using the well-known Little's Law (Little 1961) which states that the average number of customers in a queuing system can be calculated from the arrival rate λ and the average time a customer spends in the system (e.g., mean response time $\bar{T}$): $\bar{K} = \lambda \cdot \bar{T}$. Alternatively, it is possible to compute the mean number of jobs as:

$$\bar{K} = \sum_{k=1}^{\infty} k \cdot \pi_k \tag{7.2}$$

   such that π_k is the steady-state probability of the system being at the state k (e.g., consisting of k jobs), where details on deriving the steady-state probabilities will be provided in Sect. 7.3.3.
2. Average waiting time in the queue ($\bar{W}$): It demonstrates the meantime a job waits inside the queue until it is assigned to one of the available server(s). If the average time a customer waits before being served is calculated to be 10 min, this metric helps in understanding customer experience. This can be presented using the Little's Law (Little 1961) as: $\bar{W} = \frac{\bar{Q}}{\lambda}$, where $\bar{Q}$ is the average length (e.g., the mean number of jobs stored) of the queue, and λ denotes the arrival rate of jobs.
3. Utilization rate (ρ): It ranges from 0 to 1, where a value of 1 signifies an overloaded system and 0 indicates that the system is idle. For a stable queuing system, this rate is mathematically defined as $\rho = \frac{\lambda}{m \cdot \mu}$, where λ represents the arrival rate, μ is the service rate, and m is the number of servers. For instance, consider a single-server queue with an arrival rate of 8 customers per hour and a service rate of 10 customers per hour. In this scenario, the utilization rate ρ would be 0.8, meaning the server is engaged 80% of the time. It's crucial to highlight that for the system to remain stable, the arrival rate must always be less than the service capacity (e.g., $\lambda < \mu$), ensuring that the server can handle incoming jobs efficiently without becoming overwhelmed.

To accurately determine the performance metrics mentioned earlier, we need to conduct a Markovian analysis, which enables us to calculate the steady-state probabilities for the system's various states. Once we have these probabilities, we can derive all relevant performance metrics. Specifically, we can compute the average number of jobs in the system, denoted as $\bar{K}$, from the steady-state probabilities.

With the calculated $\bar{K}$ and the system parameters–arrival rate λ, service rate μ, and the number of servers m–we can apply Little's Law to derive additional key performance metrics, such as average waiting time $\bar{W}$, average queue length $\bar{Q}$, and average response time $\bar{T}$. For a deeper understanding of the Markovian analysis, please refer to Sect. 7.3.

7.2.3 *Applications of Queuing Theory*

Queuing theory has numerous practical applications across various domains. For instance, in telecommunications, network routers process incoming data packets. By employing queuing models, network engineers can effectively optimize bandwidth and minimize latency (Giambene 2014). In manufacturing, analyzing the flow of materials and machine waiting times can significantly streamline operations and alleviate bottlenecks (Papadopolous et al. 1993). In healthcare, hospitals can apply queuing theory models to enhance patient flow in emergency departments, leading to reduced waiting times and improved service delivery (Fomundam and Herrmann 2007).

Similarly, in the retail sector, evaluating customer traffic and service efficiency at checkout lines enables retailers to better manage staff schedules, ultimately boosting customer satisfaction (Nosek and Wilson 2001). In the realm of quantum computing, researchers such as those cited in Basmadjian and Paler (2023), Dobbs et al. (2021) and Paler and Basmadjian (2019) have employed queuing theory as a foundational methodology. They specifically focused on queuing systems and networks to enhance the design and efficiency of quantum circuits used for arithmetic operations like addition and multiplication.

This chapter will delve into how queuing theory models can be utilized to assess the power-performance trade-off in servers using the DVFS technology, with further details provided in Sect. 7.4. In summary, queuing theory offers a systematic framework or methodology for understanding and optimizing systems that involve waiting lines. By examining factors such as arrival and service rates, queue discipline, and server capacity, organizations can harness data-driven insights to enhance efficiency and elevate customer satisfaction. Next, we will present all the necessary information to conduct a Markovian analysis, enabling us to derive the performance metrics discussed earlier.

7.3 Markovian Analysis

7.3.1 *Markov Process*

Markovian analysis focuses on stochastic processes characterized by the Markov property, which states that the future state of a process depends only on its current state, not on the sequence of events that preceded it (e.g., history) (Dynkin 1965). Two aspects can be deduced from Markov property: (1) the next state change does not depend on previous states, and (2) the next state change does not depend on the age of the current state. This memoryless property makes Markovian processes particularly useful for various real-world applications, especially for queuing systems.

The Markov property can be mathematically presented as:

$$P(X_{t_{n+1}} \leq s_{n+1} | X_{t_n} = s_n, X_{t_{n-1}} = s_{n-1}, \ldots, X_{t_0} = s_0) = P(X_{t_{n+1}} \leq s_{n+1} | X_{t_n} = s_n) \tag{7.3}$$

where X_{t_i} is the random variable representing the stochastic process having Markovian property such that the index t_i denotes the notion of time and s_i is the corresponding state of the system.

Over time, a Markovian process may reach a steady-state distribution where the probabilities of being in each state stabilize. This is crucial for understanding long-term behavior. For instance, if a queue reaches a steady state, you can predict the average number of customers in the system. Next, we will explore various classifications of Markov processes, taking into account both spatial and temporal dimensions.

7.3.2 Markov Chains

In a Markov process, the system operates within a finite or countably infinite state space, leading to the concept of a Markov "chain" (MC). For instance, in a customer service context, possible states might include "customer waiting," "customer being served," and "customer served." Additionally, time can be categorized into discrete or continuous intervals, resulting in either a discrete-time Markov chain (DTMC) or a continuous-time Markov chain (CTMC), with specific characteristics for each type discussed further along.

In summary, a Markov chain is essentially a collection of states along with the transitions that occur between them. Next, we will delve into a detailed explanation of each type of Markov chain, beginning with discrete-time Markov chains and subsequently exploring continuous-time Markov chains. This structured approach will help clarify the distinct characteristics and applications of each type.

7.3.2.1 Discrete-Time Markov Chain

A DTMC is analyzed by examining predefined fixed time intervals or steps—such as seconds, minutes, hours, days, or years—during which transitions between states occur exclusively at these specified time steps. Each transition in a DTMC is associated with a probability, which determines the likelihood of moving from one state to another in a given time interval or step. For example, if a customer is currently being served, there may be a 90% chance they will finish being served in the next minute and a 10% chance they will require additional attention.

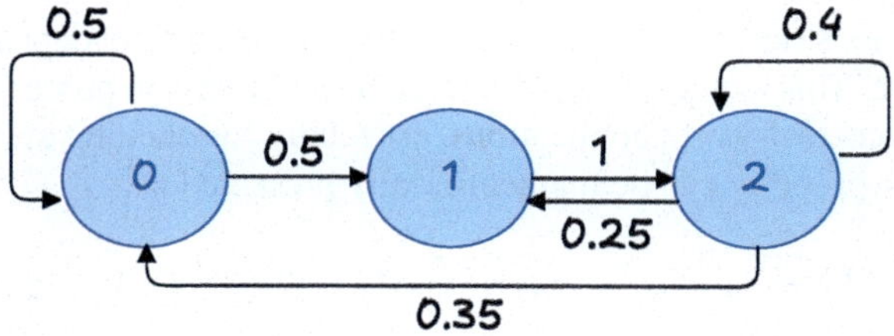

Fig. 7.5 An example of a discrete-time Markov chain (DTMC) featuring a finite state space consisting of States 0, 1, and 2, depicted by blue circles labeled with their respective numbers. Transitions between these states are illustrated by arrows, each annotated with the corresponding probability. It's important to ensure that the sum of the probabilities for all outgoing transitions from a state (for example, from State 0 to State 1 and from State 0 to itself) equals one. Additionally, self-transitions are permitted in a DTMC, as seen with State 0 and State 2

It is essential that the total of these probabilities for all outgoing transitions from any given state equals one. Notably, DTMCs permit self-transitions, allowing the system to remain in the same state during a time interval. Figure 7.5 illustrates an example of a finite state-space discrete-time Markov chain, where transitions between states occur in discrete time intervals, each associated with specific probabilities. Importantly, the total of the probabilities for all outgoing transitions from each state equals one. Additionally, self-transitions are observed at State 0 and State 2.

The transition probabilities for all states are represented in a matrix form which is called the transition probability matrix and is denoted by **P**. Each element ($\mathbf{P}_{ij}$) in the matrix represents the probability of transitioning from State i to State j. For instance, the transition probability matrix for the DTMC of Fig. 7.5 looks like this:

$$\mathbf{P} = \begin{bmatrix} 0.5 & 0.5 & 0 \\ 0 & 0 & 1 \\ 0.35 & 0.25 & 0.4 \end{bmatrix}$$

where **P** is a "stochastic" matrix such that row sums are equal to one and each element's value is within [0, 1].

7.3.2.2 Continuous-Time Markov Chain

Continuous-time Markov chains (CTMCs) extend the concept of Markov chains into continuous time, allowing transitions between states to occur at any point in time rather than at fixed intervals, which is the case for DTMCs. Similar to discrete-time Markov chains, CTMCs operate over a finite or infinite set of states. For example, a CTMC might model the states of a server being "busy", "idle", or "failed".

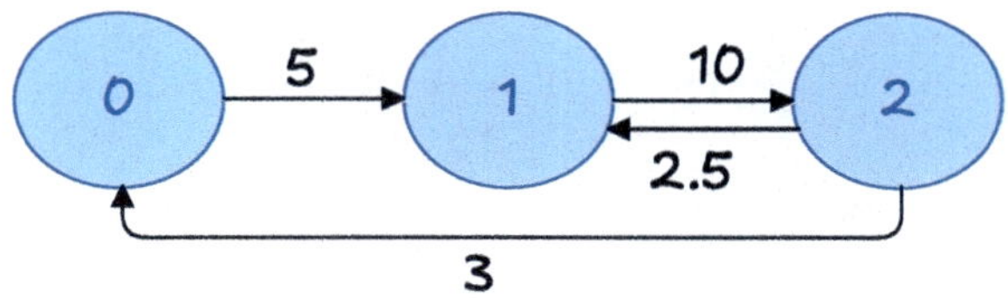

Fig. 7.6 An example of a continuous-time Markov chain (CTMC) featuring a finite state space consisting of States 0, 1, and 2, depicted by blue circles labeled with their respective numbers. Transitions between these states are illustrated by arrows, each annotated with the corresponding transition rates. It's important to note that self-transitions are not allowed in a CTMC

However, instead of transition probabilities for the case of DTMCs, CTMCs use transition rates, typically denoted as q_{ij} for moving from State i to State j. These rates indicate the instantaneous likelihood of transitioning from one state to another. In other words, the rates can be seen as the limit:

$$q_{ij} = \lim_{\Delta t \to 0} \frac{p_{ij}(t, t + \Delta t)}{\Delta t} \tag{7.4}$$

where $p_{ij}(u, v)$ is the transition probability, that the CTMC travels from State i to State j during the time interval $[u, v)$ and is given by:

$$p_{ij}(u, v) = P(X_v = j | X_u = i) \tag{7.5}$$

Figure 7.6 illustrates an example of a continuous-time Markov chain (CTMC) with a finite state space consisting of States 0, 1, and 2. When comparing this figure to Fig. 7.5, the key distinctions between CTMCs and discrete-time Markov chains (DTMCs) become evident. Firstly, transitions in a CTMC are defined by transition rates rather than probabilities. These rates indicate how quickly transitions occur between states.

Secondly, CTMCs do not permit self-transitions, meaning that once a process is in a specific state, it cannot remain there but must transition to another state. This contrasts with DTMCs, where self-transitions can occur, allowing the system to stay in the same state with a certain probability. These differences highlight the unique characteristics of CTMCs, making them particularly suitable for modeling systems where events happen continuously over time and where the dynamics of state transitions are crucial for analysis.

The transition rates can be organized into a matrix known as the infinitesimal generator matrix **Q**. The elements of this matrix describe the rates of transitions between states, with the diagonal elements given by $q_{ii} = -\sum_{j \neq i} q_{ij}$ to ensure that the total rate leaving State i is properly defined. For instance, the infinitesimal generator matrix for the CTMC of Fig. 7.6 looks like this:

$$\mathbf{Q} = \begin{bmatrix} -5 & 5 & 0 \\ 0 & -10 & 10 \\ 3 & 2.5 & -5.5 \end{bmatrix}$$

where $\mathbf{Q}$ is the generator matrix whose diagonal elements are negative row sums.

7.3.3 Closed-Form Solution

The primary goal of a Markovian analysis is to derive the unique steady-state probability vector. To ensure that such a unique solution exists, it is essential for the underlying discrete-time Markov chain (DTMC) or continuous-time Markov chain (CTMC) to be ergodic.

A Markov chain is considered ergodic, if it meets three criteria: irreducibility, positive recurrence, and aperiodicity.

1. Irreducibility: Means that every state can be accessed from any other state. For instance, envision a Markov chain representing a board game where players can traverse various positions; if the chain is irreducible, every position can be reached from any other position without restrictions. The CTMC illustrated in Fig. 7.6 is classified as irreducible because every state can be accessed from any other state through a series of transitions. For instance, State 0 can transition to State 1, and then to State 2; State 1 can transition to State 2 and subsequently back to State 0; similarly, State 2 can reach State 0 and then transition to State 1. This interconnectedness underscores the chain's irreducibility, ensuring that no state is isolated from the others.
2. Positive recurrence: Indicates that the expected time to return to any given state is finite. This suggests that if you start in a specific state, you will eventually return to it after a finite number of steps. According to the definition, any finite-state discrete-time Markov chain (DTMC) or continuous-time Markov chain (CTMC) is considered positive recurrent, since the expected return time to any state after leaving it is finite. Given that the CTMC depicted in Fig. 7.6 comprises three finite states, we can conclude that all states within this CTMC are positive recurrent.
3. Aperiodicity: Ensures that the chain does not get trapped in cycles. For example, if a player can move freely but must return to the starting position every third turn, the chain is periodic. In contrast, an aperiodic Markov chain allows transitions to any state in various steps, preventing cyclical behavior. By definition, the states of an CTMC are aperiodic which is also the case for the CTMC in Fig. 7.6.

When a Markov chain satisfies these properties, it possesses a unique stationary distribution that describes its long-term behavior (Cogburn 1984). For example, when modeling weather patterns, an ergodic Markov chain can predict that, over time, the weather will stabilize into a predictable pattern, regardless of the initial conditions.

In principle, finite DTMCs are ergodic if they are irreducible and aperiodic, whereas finite CTMCs are ergodic if they are irreducible.

Given the fact that the underlying Markov chain is ergodic, then the closed-form solution is obtained by solving $\mathbf{v} = \mathbf{v} \cdot \mathbf{P}$ for DTMCs and $\Pi \cdot \mathbf{Q} = 0$ for CTMCs, where $\mathbf{v} = [v_0, v_1, \ldots, v_n]$ and $\Pi = [\pi_0, \pi_1, \ldots, \pi_n]$ are the vectors of the state probabilities for DTMC and CTMC respectively, whereas $\mathbf{P}$ and $\mathbf{Q}$ are the corresponding transition probability matrix and infinitesimal generator matrix given in Sects. 7.3.2.1 and 7.3.2.2 respectively. Please note that the bold notations indicate the presence of vectors or matrices.

Conducting a Markovian analysis ultimately involves solving a series of linear equations (e.g., multiplication of vectors by matrices), with the complexity of this task directly related to the number of states within the system. For those seeking a deeper understanding of the queuing theory methodology and the intricacies of Markovian analysis, additional details can be found in reference (Bolch et al. 2006). This resource provides comprehensive insights into the mathematical foundations and applications of these concepts.

7.4 Appropriateness of M/M/1 Queuing Systems for Trade-Off Analysis

In this section, we illustrate how the queuing theory methods from Sect. 7.2, combined with the Markovian analysis described in Sect. 7.3, provide a solid framework for assessing the trade-off between power demand and performance in CPU-bound applications. We specifically model the CPU processing service s at a constant frequency (e.g., highest frequency $f_{\max}$) through the M/M/1-FCFS queuing system model using both continuous- and discrete-time Markov chains. After evaluating the effectiveness of this model for the use case of static frequency, we will extend it to investigate dynamic frequency scaling in Sect. 7.5.

7.4.1 Continuous-Time MC and Birth-Death Process

Figure 7.7 illustrates the CTMC of the corresponding M/M/1-FCFS queuing system. This CTMC can be visualized as a birth-death process, where births correspond to arrivals (occurring at rate λ) and deaths represent the completion of service (happening at rate μ). The balance equations derived from this model allow us to analyze various performance metrics, such as the average number of customers in the system ($\bar{K}$), average waiting time ($\bar{W}$), and the utilization of the server (ρ).

Utilizing the CTMC framework allows us to explore the queuing dynamics and performance metrics of the M/M/1-FCFS system across different arrival and service rates. We will leverage this queuing model to illustrate its effectiveness in analyzing

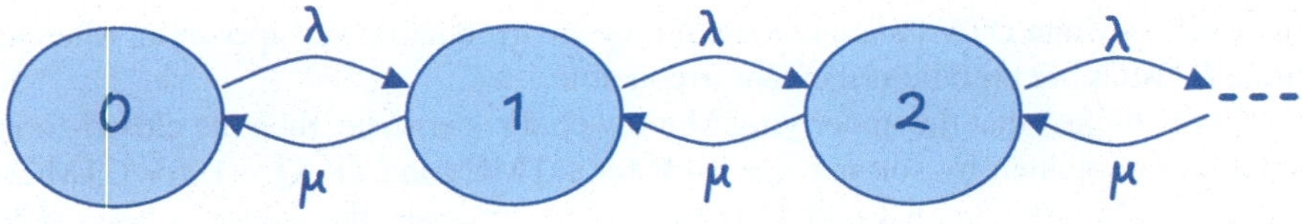

Fig. 7.7 A graphical representation of the CTMC for an M/M/1-FCFS queuing system. In this system, we have a single server queuing model, where "M" denotes that both arrival and service times follow an exponential distribution. The first "M" represents the arrival process, which occurs at a rate of λ, indicating the average number of customers arriving per time unit. The second "M" signifies that the service times are exponentially distributed with a service rate of μ, which reflects the average number of customers that can be served by the server in a time unit. In a CTMC representation, the states of the system correspond to the number of customers or jobs in the queue, including the one being served. The transition rates between these states depend on the arrival rate λ and the service rate μ. For instance, when the system is in State 1 (e.g., the total number of customers, either in service or waiting), the following transitions can occur: (1) from State 1 (e.g., n) to State 2 (e.g., $n + 1$) with a rate of λ, representing the arrival of a new customer, and (2) from State 1 (e.g., n) to State 0 (e.g., $n - 1$) with a rate of μ, indicating that the customer being served has completed its service

the power-performance trade-off. To support our exploration, we will draw insights and methodologies from the existing scientific literature on M/M/1-FCFS queuing systems.

7.4.1.1 Closed-Form Results for Performance Metrics

By applying the Markovian analysis outlined in Sect. 7.3 to the continuous-time Markov chain (CTMC) illustrated in Fig. 7.7, the literature (Bolch et al. 2006) demonstrates that the steady-state probabilities for M/M/1-FCFS queuing systems can be expressed as follows:

$$\pi_k = (1 - \rho) \cdot \rho^k \ \forall k, \text{ where } \rho = \frac{\lambda}{\mu} \tag{7.6}$$

Using Eq. (7.2) together with Eq. (7.6), we can derive the mean number of jobs in an M/M/1-FCFS queuing system as:

$$\bar{K} = \frac{\rho}{1 - \rho} \tag{7.7}$$

Using Little's Law together with Eq. (7.7), it is possible to calculate the mean response time $\bar{T}$ of the service s as:

$$\bar{T} = \frac{1}{\mu - \lambda} \tag{7.8}$$

where λ and μ denote the arrival and service rates respectively, whereas ρ is the utilization rate of the underlying system.

7.4.1.2 Closed-Form Results for Power Demand

Understanding the CPU power demand associated with each state of the CTMC enables us to calculate the average CPU power demand, denoted as $\bar{P}_{\text{CPU}}$. In this context, let $P_{\text{CPU}}^{\text{idle}}$ represent the power demand when the CPU is in an idle state, while $P_{\text{CPU}}^{\text{dynamic}}$ signifies the power demand when the CPU is operating at full capacity given by Eqs. (6.10) and (6.18) respectively.

By analyzing these power demand metrics across different states, we can derive a comprehensive picture of the CPU's power and energy requirements under various operational conditions. Based on the steady-state probabilities of M/M/1-FCFS queuing system given in Eq. (7.6), the mean power demand can be calculated as:

$$\bar{P}_{\text{CPU}} = \pi_0 \cdot P_{\text{CPU}}^{\text{idle}} + \sum_{k=1}^{\infty} \pi_k \cdot P_{\text{CPU}}^{\text{dynamic}} = \pi_0 \cdot P_{\text{CPU}}^{\text{idle}} + (1 - \pi_0) \cdot P_{\text{CPU}}^{\text{dynamic}} \tag{7.9}$$

Taking the results from the Eq. (7.6) and setting $k = 0$, we get $\pi_0 = 1 - \rho$. Thus, replacing the value of π_0 into Eq. (7.9), this leads to:

$$\bar{P}_{\text{CPU}} = (1 - \rho) \cdot P_{\text{CPU}}^{\text{idle}} + \rho \cdot P_{\text{CPU}}^{\text{dynamic}} \tag{7.10}$$

where $\rho = \frac{\lambda}{\mu} \in [0, 1[$ is the utilization rate of the CPU when running the service s, whereas $P_{\text{CPU}}^{\text{idle}}$ and $P_{\text{CPU}}^{\text{dynamic}}$ signify the idle and dynamic power demand of the CPU given in Eqs. (6.10) and (6.18) respectively.

7.4.1.3 Assessment of the M/M/1 Model

Figures 7.8 and 7.9 showcase the results from experiments[2] aimed at evaluating the M/M/1-FCFS queuing system model's effectiveness in analyzing the trade-off between power and performance. Since dynamic frequency scaling necessitates more intricate models (refer to Sect. 7.5.3), we configured the CPU to operate under the Performance governor, maintaining the highest frequency consistently. Each experiment was conducted five times, and we calculated the minimum, average, and maximum values of the results using 95% confidence intervals.

The arrival rate λ was incrementally increased from 1 job/s to 7 jobs/s, while the maximum service rate at this highest frequency was determined to be $\mu = 10$ jobs/s. A close examination of these figures reveals a clear linear relationship: as the arrival

[2] A custom-made software was developed that precisely simulates the behavior of an M/M/1-FCFS queuing system.

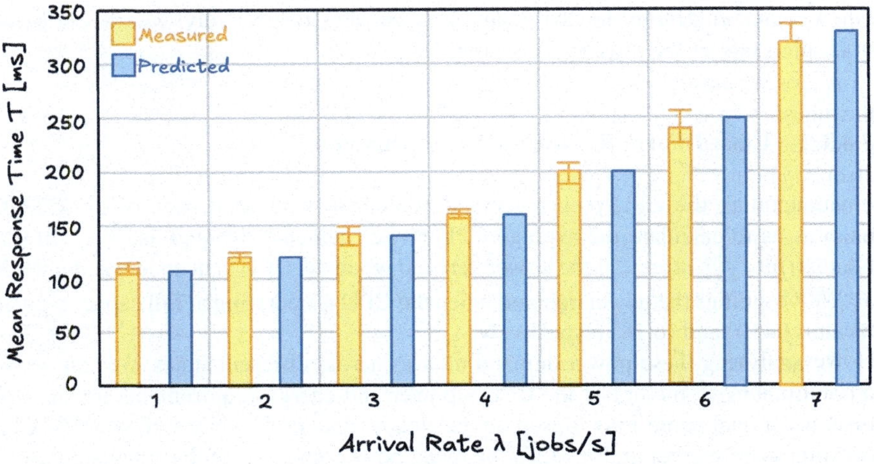

Fig. 7.8 The measured values (represented by orange bars) and predicted values (depicted by blue bars) for the mean response time $\bar{T}$ were collected while using the Performance governor, which maintains the CPU at its maximum frequency. This analysis was conducted for arrival rates $\lambda = [1, \ldots, 7]$ jobs/s with a service rate of $\mu = 10$ jobs/s. The experiments were conducted five times, and the minimum, average, and maximum values were calculated with 95% confidence intervals, which are illustrated as the boundaries of the orange bars

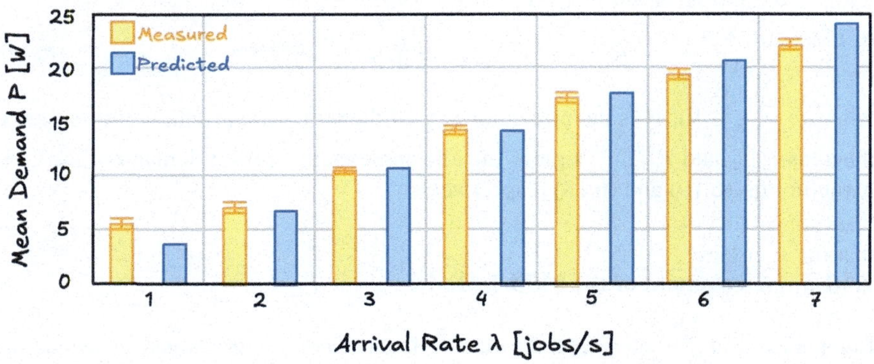

Fig. 7.9 The measured values (represented by orange bars) and predicted values (depicted by blue bars) for the average power demand $\bar{P}_{\text{CPU}}$ were collected while using the Performance governor, which maintains the CPU at its maximum frequency. This analysis was conducted for arrival rates $\lambda = [1, \ldots, 7]$ jobs/s with a service rate of $\mu = 10$ jobs/s. The experiments were conducted five times, and the minimum, average, and maximum values were calculated with 95% confidence intervals, which are illustrated as the boundaries of the orange bars

rate λ rises, both the mean response time $\bar{T}$ and the average CPU power demand $\bar{P}_{\text{CPU}}$ increase correspondingly.

A close examination of those two figures reveals noteworthy findings. In Fig. 7.8, the measured mean response time $\bar{T}_{\text{measured}}$ (e.g., orange bars) closely aligns with the predicted value $\bar{T}$ (e.g., blue bars) derived from Eq. (7.8), showing a mean relative error of about 1% and a maximum relative error of approximately 2.6%. Figure 7.9 illustrates the results of the measured mean power demand $\bar{P}_{\text{measured}}$ (e.g., orange bars) and predicted mean power demand $\bar{P}_{\text{CPU}}$ (e.g., blue bars) computed by the Eq. (7.10).

We observe that at low CPU utilization rates (below 10%), the model exhibits a mean relative error of about 9%, with a maximum relative error approaching 35%. However, for CPU utilization rates exceeding 10% (e.g., $\lambda > 1$), the mean relative error drops to around 5%, and the mean absolute error remains below 2W. These insights suggest that relatively simple Markov models of M/M/1 queuing systems can provide effective and accurate predictions for both service performance and power demand.

7.4.2 Discrete-Time MC for Time-Dependent Utilization

After assessing the M/M/1 queuing system and its continuous-time Markov chain (CTMC) model for evaluating the power-performance trade-off with a fixed CPU frequency (like the maximum setting), our next focus is to explore the development of a discrete-time Markov chain (DTMC) model.

This approach is particularly pertinent because utilization-based governors modify the CPU frequency at specific discrete intervals, such as every 10 ms. Consequently, we will introduce a DTMC model that accounts for time-dependent utilization but does not yet incorporate variations in service rates. Furthermore, we will demonstrate that the accuracy of this derived model remains consistent, regardless of the length of the time interval between two state transitions.

A DTMC model for M/M/1 queues with a constant service rate is presented in Bertsekas and Gallager (2021). In this model, let A_k represent the number of jobs arriving at the processor, and L_k denote the number of jobs departing during the time interval I_k. The authors of Bertsekas and Gallager (2021) assume that the time interval δ between state transitions is a small positive value compared to the mean service time $\bar{T}_S = \frac{1}{\mu}$ and the mean inter-arrival time $\bar{T}_A = \frac{1}{\lambda}$.

As a result, the probabilities $P[A_k > 1]$ and $P[L_k > 1]$ are extremely low and can be considered negligible. Thus, unlike the comprehensive DTMC representation of an M/M/1-FCFS queue, this model only connects neighboring states, focusing exclusively on scenarios where the number of arriving or departing jobs during the interval I_k is either 0 or 1.

Figure 7.10 illustrates the underlying DTMC, where p_{ij} is the transition probability from State i to State j such that p_{ii} denotes the self-transition probability. As

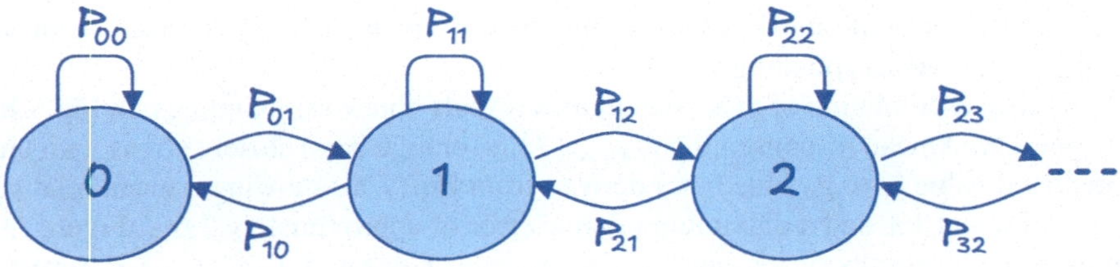

Fig. 7.10 This figure outlines a Markov chain model for the M/M/1 queuing system, focusing on discrete time intervals. At specific intervals, such as every 10 ms, the system exhibits three possible state transitions, with the exception of State 0, which has only two options: (1) remain in the current state with probability p_{ii}, indicating that during this interval, no new jobs have arrived, and the existing job is not being processed, (2) transition to the next state with probability $p_{i(i+1)}$, reflecting the arrival of a new job, and (3) move back to the previous state with probability $p_{i(i-1)}$, representing the completion of an existing job. In State 0, only Steps (1) and (2) are applicable. There is no option for Step (3) since the system is idle and cannot revert to a previous state when there are no jobs present

detailed in Bertsekas and Gallager (2021), those state transition probabilities p_{ij} can be determined as:

$$
\begin{aligned}
p_{00} &\approx P[A_k = 0] \approx 1 - \lambda\delta, & \\
p_{ii} &\approx P[A_k = 0 \wedge L_k = 0] \approx 1 - \lambda\delta - \mu\delta, & i \geq 1, \\
p_{i(i+1)} &\approx P[A_k = 1 \wedge L_k = 0] \approx \lambda\delta, & i \geq 0, \\
p_{i(i-1)} &\approx P[A_k = 0 \wedge L_k = 1] \approx \mu\delta, & i \geq 1.
\end{aligned}
\tag{7.11}
$$

It's important to remember that the DTMC model depicted in Fig. 7.10 operates under the assumption that δ is small relative to the mean service time $\bar{T}_S$ and inter-arrival time $\bar{T}_A$ of jobs. Additionally, it posits that during any time interval of length δ, only one job can either arrive or depart from the system. However, if δ is not sufficiently small, multiple jobs could indeed arrive within that interval, challenging this assumption.

Following this, we will provide a mathematical demonstration that the DTMC model is equivalent to the CTMC model discussed in Sect. 7.4.1 (refer to Fig. 7.7) for any arbitrary value of δ. The transition probability matrix $\mathbf{P}$ of the DTMC model of Fig. 7.10 is given by:

$$
\mathbf{P} = \begin{pmatrix}
1-\lambda\delta & \lambda\delta & 0 & \cdots \\
\mu\delta & 1-\lambda\delta-\mu\delta & \lambda\delta & 0 \\
0 & \mu\delta & 1-\lambda\delta-\mu\delta & \lambda\delta \\
0 & 0 & \mu\delta & 1-\lambda\delta-\mu\delta \\
\vdots & \vdots & 0 & \mu\delta
\end{pmatrix}
$$

where λ and μ are the arrival and service rates respectively, whereas δ denotes the sampling period. As previously noted, δ should be a small value relative to both the arrival and service rates so that the steady-state probabilities p_{ij} fall between 0 and 1.

As noted in Bolch et al. (2006, p. 104), a DTMC can be converted into a CTMC by transforming its transition probability matrix **P** into the generator matrix **Q** of the corresponding CTMC. This transformation involves subtracting the identity matrix **I** from **P**, leading to:

$$\mathbf{Q} = \begin{pmatrix} -\lambda\delta & \lambda\delta & 0 & \cdots \\ \mu\delta & -\lambda\delta - \mu\delta & \lambda\delta & 0 \\ 0 & \mu\delta & -\lambda\delta - \mu\delta & \lambda\delta \\ 0 & 0 & \mu\delta & -\lambda\delta - \mu\delta \\ \vdots & \vdots & 0 & \mu\delta \end{pmatrix}$$

$$= \delta \cdot \underbrace{\begin{pmatrix} -\lambda & \lambda & 0 & \cdots \\ \mu & -\lambda - \mu & \lambda & \ddots \\ 0 & \mu & -\lambda - \mu & \ddots \\ 0 & 0 & \mu & \ddots \\ \vdots & \vdots & \ddots & \ddots \end{pmatrix}}_{\mathbf{Q}'}$$

Consequently, the CTMC characterized by **Q** will exhibit the same steady-state probabilities as the DTMC represented by **P**. The steady-state probability vector $\boldsymbol{\pi}$ can be calculated by solving $\boldsymbol{\pi}\mathbf{Q} = 0$. Since $\mathbf{Q} = \delta \cdot \mathbf{Q}'$, then:

$$\begin{aligned} \boldsymbol{\pi}\mathbf{Q} = 0 &\Leftrightarrow \boldsymbol{\pi} \cdot \delta \cdot \mathbf{Q}' = 0 \\ &\Leftrightarrow \delta \cdot \boldsymbol{\pi} \cdot \mathbf{Q}' = 0 \\ &\Leftrightarrow \boldsymbol{\pi} \cdot \mathbf{Q}' = 0 \end{aligned} \tag{7.12}$$

When $\delta = 1$, **Q** serves as the generator matrix for the CTMC model outlined in Sect. 7.4.1. Consequently, the steady-state probabilities of the DTMC model align perfectly with those of the CTMC model, regardless of the value of δ.

7.5 Utilization-Based Governors with DVFS: M/M/1 Model Extension

In Sect. 7.4, we showed that the M/M/1 queuing system model effectively approximates reality when the CPU frequency for service s is set to a fixed value (such as the maximum frequency with a Performance governor). However, modern multi-core CPUs in data center servers utilize dynamic voltage and frequency scaling (DVFS) technology, which enables real-time adjustments to the frequency of CPU cores.

This hardware capability works in conjunction with utilization-based governors, like the onDemand and Conservative governors discussed in Sect. 7.1.1. These governors dynamically scale the frequency and voltage of CPU cores according to specific utilization rate thresholds.

To enrich our analysis, it's crucial to extend the M/M/1 model to include scenarios where CPU frequency is dynamically adjusted rather than being fixed at a predetermined value, as was discussed in the previous section. In this section, we will first introduce the necessary notations and outline the key assumptions underlying the corresponding methodology.

Subsequently, we'll provide a brief overview of the operational dynamics of the two utilization-based governors such as onDemand and Conservative. Following that, we will articulate the specific problem statement and delve deeper into the modeling intricacies involved in this dynamic scaling process. This comprehensive exploration will allow us to better understand how adapting CPU frequency in real-time impacts system performance and efficiency.

7.5.1 *Notations and Assumptions*

Let s represent a CPU-bound service. A host h (e.g., server) equipped with dynamic voltage and frequency scaling (DVFS) technology is dedicated solely to the service s, ensuring that no other services are utilizing the CPU resources of h: we can disregard system processes here due to their minimal impact on CPU load. It is assumed that the host h features a processor capable of operating at n different frequencies denoted as $f_0, f_1, \ldots, f_n$, where f_0 corresponds to the highest frequency and f_n to the lowest.

In the case of a multi-core processor, we further assume that the service s is specifically assigned to a designated core. Jobs for service s arrive following a Poisson process with an arrival rate of λ. These jobs are processed using a First-Come-First-Served (FCFS) scheduling discipline, with service times that are exponentially distributed. The mean service time, when the CPU operates at frequency f_x, is represented as $\frac{1}{\mu_x}$ for $0 \leq x \leq n$.

Let the service level agreement (SLA) for service s specify $\bar{T}_{\max}$, the maximum acceptable mean response time. Furthermore, the governors implemented on host h are two widely used utilization-based types: onDemand and Conservative. These

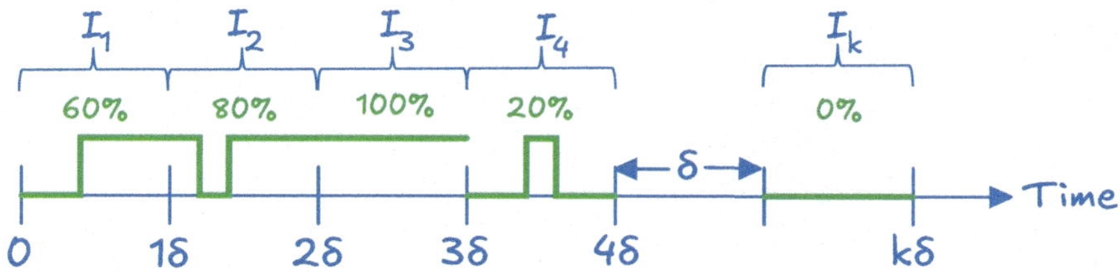

Fig. 7.11 A graphical depiction illustrates the sampling period δ with fixed time instances $k\delta$. The green line indicates the average utilization, denoted as ρ, for a specific interval I_k. This utilization is calculated by averaging the periods of idle time and full processor usage (e.g., 100%) within that interval

governors activate periodically after a specified duration, referred to as the sampling period δ. The time instances when the utilization-based governor activates are denoted as $\delta, 2\delta, \ldots, k\delta, \ldots$.

At each time instance $k\delta$, the governor evaluates the mean CPU utilization ρ_k over the interval $I_k := ((k-1)\delta; k\delta]$. It is important to note that the modeling framework requires δ to be significantly smaller than the maximum mean service time $\frac{1}{\mu_0}$. Figure 7.11 visually represents the previously discussed parameters: sampling period δ, time instances $k\delta$, and utilization rate ρ. For example, during interval I_1, the average utilization rate over δ units of time is 60%. In contrast, the utilization rate for interval I_4 drops to 20%, primarily due to a peak occurring only at the midpoint of that interval, resulting in idle time for the remainder of the duration.

7.5.2 Understanding and Modeling the Dynamics of Governors

This section illustrates the specific behaviors, rules, and decision-making processes of the two utilization-based governors understudy. We start first with the onDemand and then describe the Conservative.

7.5.2.1 onDemand

If the utilization rate ρ_{I_k} surpasses a predetermined threshold u, the CPU frequency will be adjusted to its maximum capacity, f_0, for the subsequent time interval $I_{k+1} = (k\delta; (k+1)\delta]$. Conversely, if ρ_{I_k} drops below the threshold u, the onDemand governor will reduce the CPU frequency to the next lower level for the time interval I_{k+1}. If the CPU is already operating at the lowest frequency level, f_n, it will remain at that level for the upcoming time interval I_{k+1}. This adaptive strategy ensures efficient power management by dynamically adjusting the CPU frequency based on real-time utilization metrics.

7.5.2.2 Conservative

When the utilization rate ρ_{I_k} exceeds a predefined upper threshold, denoted as u_{up}, the CPU frequency will be incrementally increased to the next level for the time interval $I_{k+1} = (k\delta; (k+1)\delta]$. This adjustment continues until the CPU reaches its maximum frequency f_0. Conversely, if ρ_{I_k} falls below a specified lower threshold, known as u_{down}, the CPU frequency will be decreased stepwise for the time interval I_{k+1}, continuing until the minimum frequency f_n is achieved. If ρ_{I_k} remains between these thresholds, the CPU frequency will hold steady for the upcoming interval.

To manage these thresholds, the Linux operating system provides access to specific system files. Those thresholds can be read and set via:

1. Up threshold: /sys/devices/system/cpu/cpu{0..n}/cpufreq/up_threshold
2. Down threshold: /sys/devices/system/cpu/cpu{0..n}/cpufreq/down_threshold.

Here, cpu{0..n} refers to the individual core number in a multi-core processor. The default settings for these thresholds are typically 80% for the up threshold and 20% for the down threshold. It's also noteworthy that the sampling period parameter δ for the Conservative governor is fixed at 80 ms and cannot be reduced further, while the onDemand governor allows for a shorter sampling period of 10 ms, enabling it to respond more quickly to changes in CPU utilization. This flexibility helps optimize performance based on specific workload demands.

7.5.3 DTMC Models

In this section, we offer a comprehensive overview and visual representation of the two unique DTMC models designed for the utilization-based onDemand and Conservative governors. We will begin by discussing the DTMC model for the onDemand governor, followed by the model for the Conservative governor.

7.5.3.1 OnDemand

This section introduces a comprehensive model of the onDemand governor, which incorporates multiple CPU frequencies based on utilization. As outlined in Sect. 7.5.2.1, if the utilization during the time interval I_k exceeds a specified threshold u, the onDemand governor adjusts the CPU frequency from f_x to the maximum frequency f_0 for the next interval I_{k+1}, where $0 < x \leq n$. In this scenario, the service rate transitions to μ_0 (e.g., fastest execution time).

Conversely, if the utilization drops below u, the governor reduces the CPU frequency from f_x to the next lower frequency f_{x+1} for the interval I_{k+1}, where $0 \leq x \leq n-1$. Should the utilization remain below u in subsequent intervals, the onDemand governor will continue to lower the frequency until reaching the lowest frequency f_n, where it stays there as long as the utilization does not exceed the threshold u.

Each frequency level f_x (where $0 \leq x \leq n$) is associated with a specific service rate μ_x. To model these varying service rates, each state of the DTMC model is denoted by i_x which represents two-dimensional information where i is the number of jobs in the system and x is the corresponding frequency of the CPU.

The transitions between States 0_x and 1_x and vice versa are taking place due to the assumption that only one job can be served or arrived within a time interval of length δ. When the system holds two or more jobs (e.g., States i_0 for $i > 1$) at the time interval $(k-1) \cdot \delta$, the system's utilization within the interval I_k reaches 100%, as the idle State 0_x cannot be reached[3] during I_k. Consequently, the service rate for the next interval I_{k+1} gets set to μ_0 (the highest rate). The transition probabilities $p_{i_0 j_0}$ for $i, j \geq 2$ are determined by Eq. (7.11), wherein μ is substituted with μ_0 in that equation.

For $i \leq 1$ at time $(k-1) \cdot \delta$, the utilization in interval I_k can vary from 0% to 100%. This variability introduces probabilities for state transitions that can lead to a reduction in the service rate. Authors in Basmadjian et al. (2016) derived the steady-state probabilities of Eq. (7.13) by deducing them from the probabilities given in Eq. (7.11). This derivation is expanded by incorporating the utilization ρ_{I_k} within the time interval $I_k := ((k-1) \cdot \delta; k \cdot \delta]$:

$$
\begin{aligned}
p_{0_n 0_n} &\approx P[A_{I_k} = 0 \wedge \rho_{I_k} \leq u] = e^{-\lambda\delta} \cdot 1 \approx 1 - \lambda\delta \\
p_{0_x 0_{x+1}} &\approx P[A_{I_k} = 0 \wedge \rho_{I_k} \leq u] = e^{-\lambda\delta} \cdot 1 \approx 1 - \lambda\delta \\
p_{0_n 1_n} &\approx P[A_{I_k} = 1 \wedge L_{I_k} = 0 \wedge \rho_{I_k} \leq u] = \underbrace{e^{-\lambda(1-u)\delta}(1 - e^{-\lambda u\delta})}_{\text{arrival after } (1-u)\delta} \approx \lambda u\delta \\
p_{0_x 1_0} &\approx P[A_{I_k} = 1 \wedge L_{I_k} = 0 \wedge \rho_{I_k} > u] = \underbrace{e^{-\lambda u\delta}(1 - e^{-\lambda(1-u)\delta})}_{\text{arrival before } (1-u)\delta} \approx \lambda(1-u)\delta \\
p_{0_x 1_{x+1}} &\approx P[A_{I_k} = 1 \wedge L_{I_k} = 0 \wedge \rho_{I_k} \leq u] = \underbrace{e^{-\lambda(1-u)\delta}(1 - e^{-\lambda u\delta})}_{\text{arrival after } (1-u)\delta} \approx \lambda u\delta \\
p_{1_n 0_n} &\approx P[A_{I_k} = 0 \wedge L_{I_k} = 1 \wedge \rho_{I_k} \leq u] = e^{-\lambda\delta}(1 - e^{-\mu_n u\delta}) \approx \mu_n u\delta \\
p_{1_x 0_{x+1}} &\approx P[A_{I_k} = 0 \wedge L_{I_k} = 1 \wedge \rho_{I_k} \leq u] = e^{-\lambda\delta}(1 - e^{-\mu_x u\delta}) \approx \mu_x u\delta \\
p_{1_x 0_0} &\approx P[A_{I_k} = 0 \wedge L_{I_k} = 1 \wedge \rho_{I_k} > u] = e^{-\lambda\delta}(1 - e^{-\mu_x\delta} - 1 + e^{-\mu_x u\delta}) \\
p_{1_x 1_0} &\approx P[A_{I_k} = 0 \wedge L_{I_k} = 0 \wedge \rho_{I_k} > u] = e^{-\lambda\delta}e^{-\mu_x\delta} \cdot 1 \approx 1 - \lambda\delta - \mu_x\delta \\
p_{1_x 2_0} &\approx P[A_{I_k} = 1 \wedge L_{I_k} = 0 \wedge \rho_{I_k} > u] = (1 - e^{-\lambda\delta})e^{-\mu_x\delta} \cdot 1 \approx \lambda\delta \\
&\text{for } i > 0: \\
p_{i_0 i_0} &\approx P[A_{I_k} = 0 \wedge L_{I_k} = 0 \wedge \rho_{I_k} > u] = e^{-\lambda\delta}e^{-\mu_0\delta} \cdot 1 \approx 1 - \lambda\delta - \mu_0\delta \\
p_{i_0 (i+1)_0} &\approx P[A_{I_k} = 1 \wedge L_{I_k} = 0 \wedge \rho_{I_k} > u] = 1 - e^{-\lambda\delta} \approx \lambda\delta \\
p_{(i+1)_0 i_0} &\approx P[A_{I_k} = 0 \wedge L_{I_k} = 1 \wedge \rho_{I_k} > u] = 1 - e^{-\mu_0\delta} \approx \mu_0\delta
\end{aligned}
\tag{7.13}
$$

[3] No two or more jobs can be serviced or arrived during the corresponding time interval.

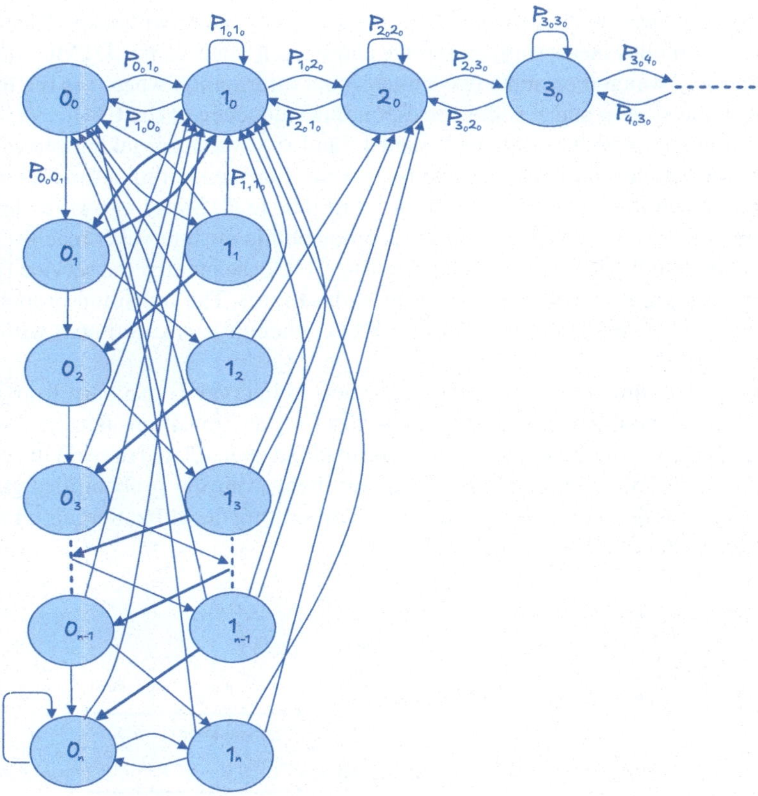

Fig. 7.12 The figure illustrates a DTMC model representing the onDemand governor's behavior. It outlines how the system moves between different states based on job arrivals and utilization rates. If the system is idle and operating at maximum frequency (e.g., f_0), it gradually reduces frequency when utilization ρ drops below the threshold u until reaching the minimum. Any State 0_x or 1_x (for $0 < x \leq n$) transitions to State 1_0 when a new task arrives and utilization exceeds a set threshold, always advancing to the highest frequency. In States i_0 for $i > 1$, the system stays at the highest frequency level independent of whether a new job arrives or an existing job departs. Please be aware that this figure does not display all state transition probabilities due to space limitations

Figure 7.12 depicts the DTMC designed to model the onDemand governor's behavior. When the system is idle (e.g., $i = 0$) and operating at the maximum frequency (e.g., $x = 0$) placed in State 0_0, if no new jobs emerge in upcoming time intervals causing the utilization ρ to fall below the defined threshold u, the system will gradually shift to lower frequencies until it reaches the minimum (e.g., $i = n$), remaining in that state thereafter.

Conversely, if the system is in State 0_0 and a new job arrives, causing the utilization ρ to surpass the set threshold u, it advances to the next state by incrementing the value of i by one while maintaining x at the maximum frequency, thus transitioning to State 1_0. Each new task arrival causes the system to move from State i_0 to State

$(i + 1)_0$ (e.g., from State 1_0 to State 2_0). The system remains in the same state i_0 as long as no new task arrives and no existing task is being processed.

Upon completing the current task, the system moves back one state from State i_0 to State $(i - 1)_0$. The scenario diverges in State 1_0 as it mirrors the behavior of other States i_0 where $i > 1$ when the utilization ρ surpasses the threshold u. However, when a job has been serviced and ρ drops below the specified threshold u, the onDemand governor opts to decrease the CPU's frequency by one level, thus transitioning from State 1_0 to State 0_1. It is important to note the transition from any State 0_x or 1_x to the State 1_0 for $0 < x \leq n$. This shift occurs solely when a new task arrives and the utilization rate ρ surpasses the threshold u within that period, consistently leaping from any existing frequency to the highest.

7.5.3.2 Conservative

This section introduces an advanced model of the Conservative governor that takes into account various CPU frequencies based on system utilization ρ. As discussed in Sect. 7.5.2.2, when the utilization surpasses a specified upper threshold u_{up} in time interval I_k, the Conservative governor adjusts the CPU frequency from f_x to the next higher feasible frequency f_{x-1} for the subsequent time interval I_{k+1}, where $0 < x \leq n$. Consequently, the service rate changes to μ_{x-1}.

If the utilization continues to exceed the upper threshold u_{up} in the following intervals, the Conservative governor progressively increases all available frequencies until reaching the highest frequency f_0. When the utilization ρ dips below a lower threshold u_{down}, the Conservative governor adjusts the CPU frequency from f_x to the subsequent lower frequency f_{x+1} for the succeeding time interval I_{k+1}, where $0 < x \leq n$.

Should the utilization persist below the lower threshold u_{down} in subsequent intervals, the Conservative governor progressively decreases all available frequencies until reaching the lowest frequency f_n. Each frequency level f_x, ranging from 0 to n, corresponds to a specific service rate μ_x, such that $\mu_x > \mu_{x+1}$. Similar to the onDemand model in Sect. 7.5.3.1, this model incorporates varying service rates by introducing new states i_x corresponding to the service rate μ_x and adapting the state transition probabilities to account for CPU utilization in interval I_k.

Figure 7.13 illustrates the proposed discrete-time Markov chain (DTMC) model for the Conservative governor. To conserve space, the figure does not include all states and their corresponding transitions. As highlighted in Sect. 7.5.2.2, the Conservative governor makes decisions based on the average processor utilization ρ measured at predefined sampling periods $\delta = 80$ ms. The use of these fixed-time instances (e.g., every 80 ms) for decision-making justifies the selection of discrete-time modeling as mentioned previously.

Similar to the DTMC of the onDemand governor, this model represents each state using two-dimensional notation, i_x, where i indicates the number of jobs in the system (with $0 \leq i \leq \infty$) and x signifies the CPU frequency, ranging from a minimum of

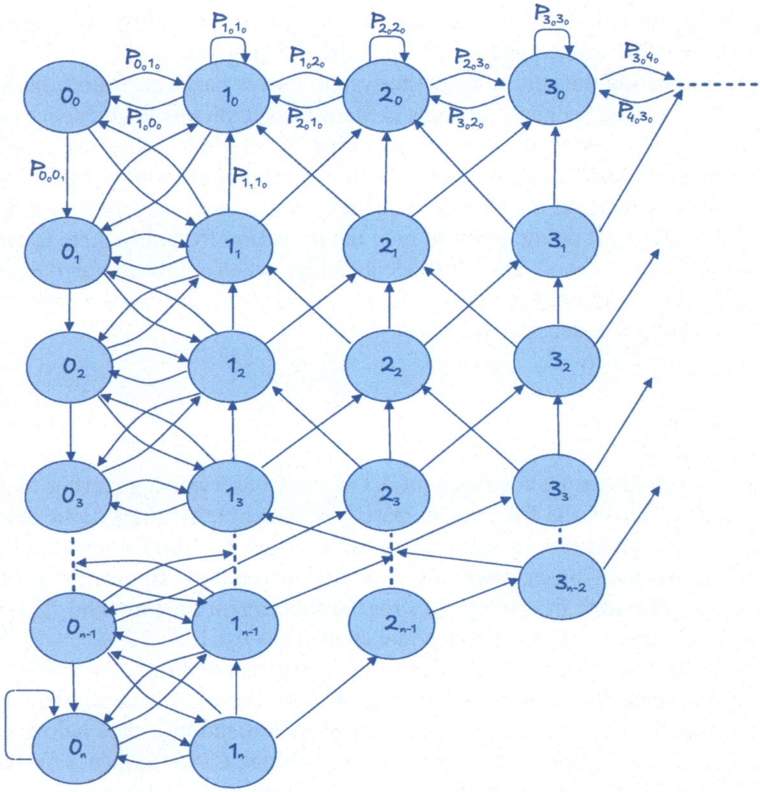

Fig. 7.13 The figure illustrates a DTMC model representing the Conservative governor's behavior. It outlines how the system moves between different states based on job arrivals and utilization rates. If the system is idle and operating at the maximum frequency (e.g., f_0), similar to the onDemand governor it gradually reduces frequency when utilization ρ drops below the down threshold u_{down} until reaching the minimum. Any State 0_x (for $0 < x \leq n$) transitions to State 1_{x+1} or 1_x or 1_{x-1} when a new task arrives and utilization drops below down threshold u_{down} or neither drops u_{down} nor it exceeds the up threshold u_{up}, or exceeds the up threshold u_{up} respectively. Any State 1_x (for $0 < x \leq n$) transitions to State 2_{x-1} upon arrival of a new job, transitions to State 0_{x+1} or 0_x or 0_{x-1} when a job is being serviced (same reason as above). Please be aware that this figure does not display all state transition probabilities due to space limitations

f_n to a maximum of f_0. When the system is idle (e.g., $i = 0$) and operating at the highest frequency (e.g., $x = 0$), it transitions to a state with one frequency increment (e.g., from 0_0 to 0_1) if no new jobs arrive.

If this condition persists over subsequent time intervals I_k, the system continues to transition from State 0_x to 0_{x+1} until it reaches the minimum frequency f_n (e.g., State 0_n), where it remains as long as no new jobs come in. These transitions occur because, during idle periods without new job arrivals, the utilization rate ρ inevitably drops below the lower threshold u_{down}.

For any state 0_x where $0 < x < n$, the arrival of a new job triggers the following transitions: (1) if the utilization rate ρ exceeds the upper threshold u_{up}, the system transitions to State 1_{x-1} (e.g., goes one frequency higher); (2) if ρ stays between the upper and lower thresholds, it moves to State 1_x (e.g., remains at the current frequency); and (3) if ρ falls below the lower threshold u_{down}, it advances to State 1_{x+1} (e.g., goes one frequency lower).

For any State i_x where $i > 1$ and $0 < x \leq n + 1 - i$, the system behaves as follows: (1) with the arrival of a new job, it transitions to the State $(i + 1)_{x-1}$; (2) if no new jobs arrive or are being serviced, it moves to State i_{x-1}; and (3) when a job is actively being serviced, it transitions to State $(i - 1)_{x-1}$. It's important to note that the system consistently shifts to one frequency higher, regardless of new job arrivals.

This is because during the time interval I_k, with more than one job present ($i > 1$), the utilization rate ρ is always above the upper threshold u_{up}. In addition to the three transitions previously described, the State 1_x also has two specific transitions when a job is being serviced: (1) it transitions to State 0_x if the utilization rate ρ remains within the upper and lower thresholds; and (2) it moves to State 0_{x+1} whenever ρ falls below the lower threshold u_{down}.

When the system is idle and functioning at the highest frequency (e.g., $x = 0$), the arrival of a new job triggers a transition from State i_0 to the State $(i + 1)_0$. It will remain in this new state if no new job arrives and no existing job is processed; otherwise, if a job is being processed, then it will revert back to State i_0.

It's important to highlight that there are no transitions between States i_x and $(i + 1)_x$ when a new job arrives for $i, x > 0$. This is because the arrival of the new job keeps the system fully occupied during that sampling period δ, causing the average utilization rate ρ to exceed the upper threshold u_{up}.

Let $p_{i_x j_x}$ be the state transition probability from state i_x to j_x where $0 \leq i, j \leq \infty$ and $0 \leq x \leq n$. The authors in Basmadjian and de Meer (2018) derived the steady-state probabilities of Eq. (7.14) as:

$$
\begin{aligned}
&\textbf{State } 0_0: \\
p_{0_0 1_0} &\approx P[A_{I_k} = 1 \wedge L_{I_k} = 0 \wedge \rho_{I_k} > u_{up}] = 1 - \lambda(1 - u_{up})\delta - (1 - \lambda\delta) \approx \lambda u_{up}\delta \\
p_{0_0 1_1} &\approx P[A_{I_k} = 1 \wedge L_{I_k} = 0 \wedge \rho_{I_k} < u_{down}] = \underbrace{e^{-\lambda(u_{up})\delta}(1 - e^{-\lambda(1-u_{up})\delta})}_{\text{arrival after } (u_{up})\delta} \approx \lambda(1 - u_{up})\delta \\
p_{0_0 0_1} &\approx P[A_{I_k} = 0 \wedge \rho_{I_k} < u_{down}] = e^{-\lambda\delta} \cdot 1 \approx 1 - \lambda\delta \\
&\textbf{State } 1_0: \\
p_{1_0 1_0} &\approx P[A_{I_k} = 0 \wedge L_{I_k} = 0 \wedge \rho_{I_k} > u_{up}] = e^{-\lambda\delta}.(1 - e^{-\mu_0\delta}) \approx 1 - \lambda\delta - \mu_0\delta \\
p_{1_0 0_0} &\approx P[A_{I_k} = 0 \wedge L_{I_k} = 1 \wedge \rho_{I_k} > u_{up}] \approx \mu_0 u_{up}\delta \\
p_{1_0 0_1} &\approx P[A_{I_k} = 0 \wedge L_{I_k} = 1 \wedge \rho_{I_k} \leq u_{down}] = e^{-\lambda\delta}.(1 - e^{-\mu_0(u_{down}\delta}) \approx \mu_0 u_{down}\delta \\
p_{1_0 2_0} &\approx P[A_{I_k} = 1 \wedge L_{I_k} = 0 \wedge \rho_{I_k} > u_{up}] = (1 - e^{-\lambda\delta}).e^{-\mu_0\delta} \approx \lambda\delta \\
&\textbf{State } 0_n: \\
p_{0_n 0_n} &\approx P[A_{I_k} = 0 \wedge \rho_{I_k} \leq u_{up}] = e^{-\lambda\delta} \cdot 1 \approx 1 - \lambda\delta \\
p_{0_n 1_n} &\approx P[A_{I_k} = 1 \wedge L_{I_k} = 0 \wedge \rho_{I_k} \leq u_{up}] = \underbrace{e^{-\lambda(1-u_{up})\delta}(1 - e^{-\lambda u_{up}\delta})}_{\text{arrival after } (1-u_{up})\delta} \approx \lambda u_{up}\delta \\
p_{0_n 1_{n-1}} &\approx P[A_{I_k} = 1 \wedge L_{I_k} = 0 \wedge \rho_{I_k} > u_{up}] = \underbrace{e^{-\lambda(1-u_{down})\delta}(1 - e^{-\lambda u_{down}\delta})}_{\text{arrival before } (u_{down})\delta} \approx \lambda u_{down}\delta \\
&\textbf{State } 1_n: \\
p_{1_n 0_n} &\approx P[A_{I_k} = 0 \wedge L_{I_k} = 1 \wedge \rho_{I_k} \leq u_{up}] = e^{-\lambda\delta}.(1 - e^{-\mu_n(u_{down}\delta}) \approx \mu_n u_{down}\delta \\
p_{1_n 0_{n-1}} &\approx P[A_{I_k} = 0 \wedge L_{I_k} = 1 \wedge \rho_{I_k} > u_{up}] = e^{-\lambda\delta}.(1 - e^{-\mu_n(u_{up}\delta}) \approx \mu_n u_{up}\delta \\
p_{1_n 1_{n-1}} &\approx P[A_{I_k} = 0 \wedge L_{I_k} = 0 \wedge \rho_{I_k} > u_{up}] = e^{-\lambda\delta}.e^{-\mu_x\delta} \approx 1 - \lambda\delta - \mu_n\delta \\
p_{1_n 2_{n-1}} &\approx P[A_{I_k} = 1 \wedge L_{I_k} = 0 \wedge \rho_{I_k} > u_{up}] = (1 - e^{-\lambda\delta}).e^{-\mu_x\delta} \approx \lambda\delta \\
&\textbf{State } k_0: \\
p_{k_0 (k-1)_0} &\approx P[A_{I_k} = 0 \wedge L_{I_k} = 1 \wedge \rho_{I_k} > u_{up}] \approx \mu_0 u_{up}\delta \\
p_{k_0 k_0} &\approx P[A_{I_k} = 0 \wedge L_{I_k} = 0 \wedge \rho_{I_k} > u_{up}] \approx 1 - \mu_0 u_{up}\delta \\
&\textbf{State } i_0: \quad 1 < i < k \\
p_{i_0 i_0} &\approx P[A_{I_k} = 0 \wedge L_{I_k} = 0 \wedge \rho_{I_k} > u_{up}] \approx 1 - \lambda\delta - \mu_0 u_{up}\delta \\
p_{i_0 (i-1)_0} &\approx P[A_{I_k} = 0 \wedge L_{I_k} = 1 \wedge \rho_{I_k} > u_{up}] \approx \mu_0 u_{up}\delta \\
p_{i_0 (i+1)_0} &\approx P[A_{I_k} = 1 \wedge L_{I_k} = 0 \wedge \rho_{I_k} > u_{up}] = (1 - e^{-\lambda\delta}).e^{-\mu_0\delta} \approx \lambda\delta \\
&\textbf{State } 0_x; 0 < x < n \\
p_{0_x 0_{x+1}} &\approx P[A_{I_k} = 0 \wedge \rho_{I_k} < u_{down}] = e^{-\lambda\delta} \cdot 1 \approx 1 - \lambda\delta \\
p_{0_x 1_x} &\approx P[A_{I_k} = 1 \wedge L_{I_k} = 0 \wedge u_{down} < \rho_{I_k} \leq u_{up}] = \\
&\quad e^{-\lambda u_{down}\delta}.e^{-\lambda(1-u_{up})\delta}.(1 - e^{-\lambda(u_{up}-u_{down})\delta}) \approx \lambda(u_{up} - u_{down})\delta \\
p_{0_x 1_{x-1}} &\approx P[A_{I_k} = 1 \wedge L_{I_k} = 0 \wedge \rho_{I_k} > u_{up}] = \underbrace{e^{-\lambda(1-u_{down})\delta}(1 - e^{-\lambda u_{down}\delta})}_{\text{arrival before } (u_{down})\delta} \approx \lambda u_{down}\delta \\
p_{0_x 1_{x+1}} &\approx P[A_{I_k} = 1 \wedge L_{I_k} = 0 \wedge \rho_{I_k} < u_{down}] = \underbrace{e^{-\lambda(u_{up})\delta}(1 - e^{-\lambda(1-u_{up})\delta})}_{\text{arrival after } (u_{up})\delta} \approx \lambda(1 - u_{up})\delta \\
&\textbf{State } 1_x; 0 < x < n \\
p_{1_x 0_x} &\approx P[A_{I_k} = 0 \wedge L_{I_k} = 1 \wedge u_{down} < \rho_{I_k} < u_{up}] \approx \mu_x\delta(1 - 2u_{down}) \\
p_{1_x 0_{x+1}} &\approx P[A_{I_k} = 0 \wedge L_{I_k} = 1 \wedge \rho_{I_k} \leq u_{down}] \approx \mu_x u_{down}\delta \\
p_{1_x 0_{x-1}} &\approx P[A_{I_k} = 0 \wedge L_{I_k} = 1 \wedge \rho_{I_k} > u_{up}] = e^{-\lambda\delta}.(1 - e^{-\mu_x(u_{down}\delta}) \approx \mu_x u_{down}\delta \\
p_{1_x 1_{x-1}} &\approx P[A_{I_k} = 0 \wedge L_{I_k} = 0 \wedge \rho_{I_k} > u_{up}] = e^{-\lambda\delta}.e^{-\mu_x\delta} \approx 1 - \lambda\delta - \mu_x\delta \\
p_{1_x 2_{x-1}} &\approx P[A_{I_k} = 1 \wedge L_{I_k} = 0 \wedge \rho_{I_k} > u_{up}] = (1 - e^{-\lambda\delta}).e^{-\mu_x\delta}) \approx \lambda\delta \\
&\textbf{Remaining states} \\
p_{i_x i-1_{x-1}} &\approx P[A_{I_k} = 0 \wedge L_{I_k} = 1 \wedge \rho_{I_k} > u_{up}] \approx \mu_x u_{down}\delta \\
p_{i_x i+1_{x-1}} &\approx P[A_{I_k} = 1 \wedge L_{I_k} = 0 \wedge \rho_{I_k} > u_{up}] \approx \lambda\delta + \mu_x u_{up}\delta \\
p_{i_x i_{x-1}} &\approx P[A_{I_k} = 0 \wedge L_{I_k} = 0 \wedge \rho_{I_k} > u_{up}] \approx 1 - \lambda\delta - \mu_x\delta
\end{aligned}
\tag{7.14}
$$

7.6 Summary

Hardware manufacturers achieve a balance between power demand and performance using dynamic voltage and frequency scaling (DVFS) technology. This chapter provides a comprehensive overview of DVFS, exploring its core functionality and examining various governors, including onDemand, Conservative, Performance, Powersave, and Userspace. Each governor's distinct rules for dynamically adjusting processor frequency are outlined in detail. Additionally, the chapter presents mathematical metrics to quantify both performance and power requirements, offering readers a holistic understanding of the trade-off involved.

The second part of this chapter delves into queuing theory, recognized as the most effective method for analyzing the performance of computing systems. It begins by outlining key modeling parameters, followed by an exploration of the three primary queuing systems: single-server, multi-server, and infinite-server systems. The section concludes with a discussion of the various performance metrics used in queuing systems, providing a comprehensive understanding of their efficiency and effectiveness.

The third part of this chapter focuses on providing a thorough overview of Markovian analysis to achieve precise insights into the power-performance trade-off. It begins by defining what a Markov process entails, followed by detailed graphical representations of the two primary types of Markov chains: discrete-time Markov chains (DTMC) and continuous-time Markov chains (CTMC). The section wraps up by outlining the key methods for conducting Markovian analysis using both DTMC and CTMC models, equipping readers with the tools needed for effective analysis.

The fourth section focuses on establishing the suitability of the single-server M/M/1 queuing system for analyzing the trade-off between performance and power. This analysis assumes that the processor operates at a constant frequency, typically at its maximum, without incorporating any dynamic scaling techniques. The evaluation reveals that this queuing model, under the constant frequency assumption, yields highly accurate results when compared to real-world performance and power demand metrics.

The final section of this chapter introduces discrete-time Markov chain models for the utilization-based governors, specifically the onDemand and Conservative approaches, while accounting for dynamic frequency and voltage scaling. Following the introduction of key mathematical notations and definitions, these were employed to model the behaviors of both governors, enhancing readers' understanding of their dynamics. The section concludes with a comprehensive explanation of the steady-state probabilities associated with the DTMC models representing the onDemand and Conservative governors.

References

Attia KM, El-Hosseini MA, Ali HA (2017) Dynamic power management techniques in multi-core architectures: A survey study. Ain Shams Eng J 8(3):445–456

Basmadjian R, de Meer H (2018) Modelling and analysing conservative governor of dvfs-enabled processors. In: Proceedings of the ninth international conference on future energy systems, pp 519–525

Basmadjian R, Niedermeier F, de Meer H (2016) Modelling performance and power consumption of utilisation-based dvfs using m/m/1 queues. In: Proceedings of the seventh international conference on future energy systems, pp 1–11

Basmadjian R, Paler A (2023) Queuing theory models for (fault-tolerant) quantum circuits: Analysis and optimization. In: Quantum computing: circuits, systems, automation and applications, pp 141–155. Springer

Begum R, Werner D, Hempstead M, Prasad G, Challen G (2015) Energy-performance trade-offs on energy-constrained devices with multi-component dvfs. In: 2015 IEEE international symposium on workload characterization, pp 34–43. IEEE

Bertsekas D, Gallager R (2021) Data networks. Athena Scientific

Bolch G, Greiner S, De Meer H, Trivedi KS (2006) Queueing networks and Markov chains: modeling and performance evaluation with computer science applications. Wiley

Brodowski D, Golde N, Wysocki RJ, Kumar V (2013) Cpu frequency and voltage scaling code in the linux (tm) kernel. Linux Kernel Documentation, 66

Carroll A, Heiser G (2014) Unifying dvfs and offlining in mobile multicores. In: 2014 IEEE 19th real-time and embedded technology and applications symposium (RTAS), pp 287–296. IEEE

Cogburn R (1984) The ergodic theory of markov chains in random environments. Zeitschrift für Wahrscheinlichkeitstheorie und verwandte Gebiete 66(1):109–128

Dobbs EE, Basmadjian R, Paler A, Friedman JS (2021) Fast swapping in a quantum multiplier modelled as a queuing network. In: International conference on reversible computation, pp 256–265. Springer

Dynkin EB (1965) Markov processes. Springer

Eyerman S, Eeckhout L (2011) Fine-grained dvfs using on-chip regulators. ACM Trans Arch Code Optim (TACO) 8(1):1–24

Fomundam S, Herrmann JW (2007) A survey of queuing theory applications in healthcare

Giambene G (2014) Queuing theory and telecommunications, vol 585. Springer

Haghighi AM, Mishev DP (2008) Queuing models in industry and business. Nova Publishers

Halimi J-P, Pradelle B, Guermouche A, Triquenaux N, Laurent A, Beyler JC, Jalby W (2013) Reactive dvfs control for multicore processors. In: 2013 IEEE international conference on green computing and communications and IEEE internet of things and IEEE cyber, physical and social computing, pp 102–109. IEEE

Hanumaiah V, Vrudhula S (2012) Energy-efficient operation of multicore processors by dvfs, task migration, and active cooling. IEEE Trans Comput 63(2):349–360

Kim JM, Kim YG, Chung SW (2013) Stabilizing cpu frequency and voltage for temperature-aware dvfs in mobile devices. IEEE Trans Comput 64(1):286–292

Krzywda J, Ali-Eldin A, Carlson TE, Östberg P-O, Elmroth E (2018) Power-performance tradeoffs in data center servers: Dvfs, cpu pinning, horizontal, and vertical scaling. Future Gener Comput Syst 81:114–128

Kuehn PJ, Mashaly M (2019) Dvfs-power management and performance engineering of data center server clusters. In: 2019 15th annual conference on wireless on-demand network systems and services (WONS), pp 91–98. IEEE

Little JD (1961) A proof for the queuing formula: L= λ w. Oper Res 9(3):383–387

Nosek RA, Wilson JP (2001) Queuing theory and customer satisfaction: a review of terminology, trends, and applications to pharmacy practice. Hosp Pharmacy 36(3):275–279

Paler A, Basmadjian R (2019) Clifford gate optimisation and t gate scheduling: using queueing models for topological assemblies. In: 2019 IEEE/ACM international symposium on nanoscale architectures (NANOARCH), pp 1–5. IEEE
Pallipadi V, Starikovskiy A (2006) The ondemand governor. Proceedings of the linux symposium 2:215–230
Papadopolous H, Heavey C, Browne J (1993) Queueing theory in manufacturing systems analysis and design. Springer Science & Business Media
Samual J, Hussin M (2018) Analysis of dvfs technique for efficient-energy management in cloud data center. Int J Eng & Technol 7(4.31):516–520
Sundarapandian V (2009) Probability, statistics and queuing theory. PHI Learning Pvt, Ltd
Wang S, Qian Z, Yuan J, You I (2017) A dvfs based energy-efficient tasks scheduling in a data center. IEEE Access 5:13090–13102

Part IV
Concluding Remarks and Emerging Opportunities

Chapter 8
Navigating the Data Center Dilemma in a Rapidly Evolving Digital World

8.1 Short and Mid-Term Challenges

The interplay between rapid technological advancement and the energy sector's ability to handle these escalating demands is pivotal for the future of data centers and the broader digital economy. A recent Bloomberg article published in August 2024 (Saul 2024) delved into the situation involving Dominion Energy (Jaye 2024), the primary electric utility in Northern Virginia, often referred to as "Data Center Alley", which is recognized as the largest global hub for data centers.

The challenges outlined in the article extend beyond mere power supply issues; they encompass vital themes such as economic growth, infrastructure development, and sustainability. These factors are intertwined, as the growing reliance on data centers for businesses requires that energy utilities/providers not only meet current demands but also anticipate future needs. As organizations navigate these complexities, fostering collaboration between technology companies and energy utilities/providers becomes crucial for establishing a more robust and resilient infrastructure.

Next, we will explore the key challenges highlighted in the Bloomberg article. These issues are not only pertinent to Dominion Energy but also resonate with electric utilities worldwide that serve regions identified as major data center hubs like Northern Virginia. Addressing these challenges will be essential for supporting the expanding digital economy and ensuring that energy sources align with the fast-paced evolution of technology.

R. Basmadjian, *Demand-Side Energy and Power Management in Data Centers*,
Lecture Notes in Energy 1, https://doi.org/10.1007/978-3-031-85956-4_8

8.1.1 Demand Surge and Waiting Times for Power Hookups

The ongoing digital transformation across various sectors is generating an extraordinary surge in demand for data processing and storage solutions. Companies such as Amazon Web Services (AWS) and Microsoft Azure are significantly expanding their data center infrastructures to accommodate the soaring need for cloud services. In 2022, AWS reported an impressive 37% year-over-year revenue growth, a clear indication of the escalating requirements for additional data center capacity. The high-performance computing (HPC) data center industry has also been experiencing remarkable growth, as evidenced by the latest Top500 list from June 2024 (Top500 2024), which ranks the world's two most powerful supercomputers, Frontier and Aurora, both located in the U.S. These impressive machines have a power demand of 23 MW and 38.7 MW, respectively.

This rapid increase in demand for data centers has outstripped the ability of energy utilities/providers to furnish the necessary electrical connections. The aforementioned article highlighted the situation with Dominion Energy in Northern Virginia, which revealed that the time required to connect large data centers (those with over 100 MW of demand) to the electric grid has been extended by one to three years. For example, a new data center planned by Google in Northern Virginia has faced delays in securing power due to existing infrastructure constraints. This dramatic rise in connection times, driven by a flood of requests, can lead to total waiting periods of up to seven years!

This scenario underscores a critical challenge for data center operators (DCO) and tech companies alike, as they grapple with the intersection of rapid technological advancement and the limitations of existing power infrastructure. The extended waiting times not only impact operational timelines but also can stall construction, delay project launches, and lead to significant financial losses, estimated in millions for companies that rely on timely infrastructure support posing potential risks to business growth and investment strategies in an increasingly competitive digital landscape.

8.1.2 Infrastructure Strain and Regional Impact

Dominion Energy has openly acknowledged the significant challenges it faces in meeting the rising power demands of the burgeoning data center industry. In 2021, the utility reported a remarkable surge in electrical service requests from data centers, receiving applications that collectively sought over 10 gigawatts of capacity—an amount that could power millions of households. As Virginia has emerged as a prime location for data center expansion, Dominion Energy has found itself grappling with the pressing need to upgrade its electrical grids to keep pace with this rapid growth.

In particular, the concentration of data centers in areas like Loudoun County—commonly known as "Data Center Alley"—has intensified the strain on local power infrastructure. This region is now home to more than 30 significant data centers, creating a scenario where the power demands are not only substantial but also concentrated. Such a high density of facilities raises legitimate concerns about the reliability and stability of the power supply.

Delays in establishing power connections can have serious repercussions, potentially causing businesses to reassess their investment strategies. Companies might opt to relocate to states or regions with more resilient and adequately developed infrastructure, where power supply can more reliably meet their needs. This situation highlights the urgent need for both energy utilities/providers (e.g., Dominion Energy) and other stakeholders to enhance their infrastructure planning and investment, ensuring that the region can sustain its status as a leading hub for data center development while fostering economic growth.

8.1.3 Sustainability Concerns and Planning Solutions

The dependence on fossil fuels for energy generation poses a significant challenge in today's push for sustainable practices. Although some utilities worldwide, like Dominion Energy and many others, are actively investing in renewable energy sources, their current energy portfolio still relies heavily on natural gas and coal. This dependency raises alarms for companies striving to meet ambitious sustainability targets. For instance, Meta (formerly Facebook) has pledged to operate on 100% renewable energy (Acun et al. 2023). However, any delays in accessing green energy can jeopardize these commitments, creating a disconnect between corporate sustainability goals and the available energy infrastructure.

To combat these challenges, several DCOs and energy utilities/providers are proactively exploring innovative solutions. A noteworthy example is Microsoft, which has taken the initiative to build its own renewable energy facilities in collaboration with utility companies. This strategic move ensures a consistent and reliable supply of clean energy for its data centers, aligning its operations with its sustainability commitments.

Furthermore, many energy utilities (e.g., Dominion Energy) have proposed an ambitious plan to upgrade their energy infrastructure. This includes significant investments in constructing new substations and enhancing transmission lines, aimed at better accommodating the escalating demands of future data centers. By expanding capacity and integrating more renewable sources, those utilities seek to create a more resilient energy system that not only meets current needs but also supports the long-term sustainability goals of its corporate partners.

8.2 Emerging Technologies to Address the Challenges

In this section, based on a personal point of view and assessment, the technologies and trends will be presented that can be used to partially or totally get rid of the challenges mentioned in Sect. 8.1.

8.2.1 Renewable Energy Sources

Renewable energy sources are energy resources that are replenished naturally and can be harnessed without depleting the Earth's resources. Here's a closer look at some of the most prominent types:

- Solar energy: This is energy harnessed from the sun using solar panels (photovoltaics) or solar thermal systems. For example, a solar farm can convert sunlight into electricity, providing sustainable power to thousands of homes or tens of medium-sized data centers. Additionally, solar energy reduces reliance on fossil fuels and can significantly decrease carbon emissions.
- Wind energy: Wind turbines convert kinetic energy from wind into electricity. Offshore wind farms, like those found off the coasts of Denmark, have become particularly popular, providing clean energy while minimizing land usage. A notable example is the Hornsea One project (U. I.-c. A. D. 2014) in the UK, which is one of the largest offshore wind farms in the world.
- Hydropower: This source uses flowing water to generate electricity, typically through dams. The Hoover Dam in the United States, for instance, produces a substantial amount of the electricity consumed in Nevada and Arizona (Bonsor 2001). Hydropower is efficient and can be a stable energy source, although it can impact local ecosystems.
- Biomass: This involves using organic materials, such as plant and animal waste, to produce energy. Biomass can be converted to biofuels like ethanol or biodiesel, which can power vehicles. A practical example is the use of agricultural waste to produce bioenergy, helping to reduce landfill waste while generating power.
- Geothermal energy: This source taps into the heat from the Earth's core to generate electricity or provide direct heating. Geothermal plants, like the Geysers in California (Sanyal and Enedy 2011), utilize steam from underground reservoirs to generate electricity, providing a reliable and consistent energy source.
- Tidal and wave energy: These emerging technologies harness the energy produced by ocean tides and waves. Tidal energy projects like the Sihwa Lake tidal power station in South Korea demonstrate how tidal currents can be exploited to generate electricity (Bae et al. 2010).

Each of these sources plays a critical role in the transition to a more sustainable energy future, helping to reduce greenhouse gas emissions and dependence on fossil fuels. The continued innovation and investment in renewable technologies will be key to addressing global energy demands while mitigating climate change. In the realm of data centers, the most promising technology is the implementation of solar energy via on-site photovoltaic (PV) panel installations (Pierson et al. 2019; Temiz and Dincer 2022). On sunny days, these panels can generate enough electricity to fully power the data center, provided their installed capacity meets the demand. As a result, data centers outfitted with solar panels can significantly alleviate the burden on energy utilities and grid operators, particularly during peak electricity usage periods. This not only enhances energy efficiency but also supports the broader transition to sustainable energy solutions.

In the realm of high-performance computing data centers, the Green500 list (Green500 2024) employs a crucial energy efficiency metric, calculated as the ratio of maximum performance (measured in GFlop/s) to maximum power demand (in Watts). This metric highlights how effectively supercomputers utilize energy while delivering high performance. Notably, the Frontier supercomputer, while being the most powerful with its unparalleled capabilities, ranks 13th on the Green500 list (published on June 2024) for energy efficiency. In contrast, the Jedi supercomputer at Forschungszentrum Jülich (FZJ) in Germany claims the top spot, showcasing an exceptional balance between performance and energy usage. This comparison underscores the ongoing effort to prioritize sustainability in high-performance computing alongside sheer computational power.

8.2.2 Edge Computing

Edge computing is a transformative technology that shifts data processing closer to the source of data generation, rather than relying on centralized data centers. This approach enhances energy efficiency and performance as well as reduces latency, making it particularly valuable for applications requiring real-time data analysis. Here are some key aspects and examples of edge computing:

- Lower latency and faster processing: By processing data at the edge, information can be analyzed in real-time without the delays associated with transmitting data to distant servers in data centers. For example, in autonomous vehicles, edge computing enables immediate processing of sensor data to make split-second driving decisions. Furthermore, localized data processing, which allows for quicker decision-making, can reduce the energy consumption of devices waiting for cloud responses. In smart manufacturing, machines can operate more efficiently by making real-time adjustments based on locally processed data, leading to less energy waste.
- Reduced data transmission and bandwidth efficiency: By processing data closer to where it's generated, edge computing minimizes the need to send large volumes of data to centralized data centers, conserving bandwidth and energy consumption for transmission. In smart cities, traffic cameras can analyze video feeds locally to detect congestion or accidents, only sending relevant data to the servers of the data

centers, which minimizes network strain and reduces the energy costs associated with transmitting that data over long distances.
- Optimized resource utilization: Edge devices can perform specific tasks tailored to local needs, which can lead to more efficient use of resources (Basmadjian et al. 2022). In smart buildings, for example, edge computing (e.g., smart thermostats) can optimize heating, cooling, and lighting based on real-time occupancy data, leading to significant energy savings. Moreover, the IoT devices utilized in edge computing are inherently designed to be highly energy-efficient, especially when contrasted with the general-purpose servers typically found in data centers (Boubouh et al. 2023a, b, c). These edge devices, such as mobile phones, sensors, gateways, and specialized computing units like IoT devices, are optimized for low power demand and energy consumption and are often equipped with advanced energy-saving technologies. For instance, many of these devices can operate on minimal power while performing tasks such as real-time data analysis and processing. In contrast, traditional data center servers are built for high-performance computing and often consume significantly more energy due to their need to handle large-scale workloads and support numerous applications simultaneously.
- Distributed load management and improved reliability: By distributing computing tasks across many edge devices, the overall load on central servers is reduced, leading to lower energy consumption at data centers. This can be particularly beneficial during peak usage times, where centralized data centers may struggle with high energy demands. Additionally, local processing can continue even if the connection to the cloud is temporarily lost. For instance, in industrial IoT applications, machines can run seamlessly and make decisions based on local data, ensuring consistent operation despite connectivity issues.

In practice, we are seeing more and more use cases in the industry leveraging edge computing such as retail, manufacturing, and telecommunications. For instance, in retail, smart shelves equipped with sensors can analyze inventory levels in real-time, optimizing restocking processes and improving customer experience (Kong et al. 2022). In manufacturing, predictive maintenance can be performed at the edge, with machines analyzing their own performance to predict failures before they occur (Yu et al. 2022). Finally, in telecommunications, 5G networks rely on edge computing to deliver high-speed, low-latency services for applications like augmented reality and virtual reality (Liu et al. 2020).

In summary, edge computing not only enhances performance and reduces latency but also leads to significant energy savings by optimizing resource utilization, minimizing data transmission, and enabling smarter, localized decision-making across various applications. This makes it a compelling choice for organizations aiming to improve their energy efficiency and sustainability efforts.

8.2.3 Advanced RISC Machine

ARM-based technology has become increasingly popular for processing due to its energy efficiency, performance capabilities, and flexibility. Next, we give a detailed look at ARM technology, its architecture, and its applications in various fields.

8.2.3.1 Architecture Overview

Advanced RISC machine (ARM) architecture is based on reduced instruction set computing (RISC) principles, which means it uses a smaller set of simple instructions. To this end, ARM processors use a variety of instruction sets, including ARM, Thumb (16-bit instructions), and ARMv8-A (supporting 64-bit processing). The ability to switch between these sets allows developers to optimize for performance or efficiency as needed. This leads to more efficient processing, lower power demand and energy consumption, and reduced heat generation. The ARM architecture boasts exceptional scalability, making it suitable for a diverse range of devices, from tiny sensors and IoT gadgets to robust servers and supercomputers. Specifically, the ARM Cortex-M series caters to microcontrollers in embedded systems, while the Cortex-A series is tailored for applications demanding higher performance levels.

8.2.3.2 Energy Efficiency

ARM processors are specifically engineered for low power demand, which positions them as the perfect choice for battery-operated devices. For example, ARM-based chips are widely utilized in Internet of Things (IoT) gadgets like smart thermostats, as well as in smartphones and tablets where energy efficiency is crucial for extending battery life. Moreover, ARM processors frequently incorporate dynamic voltage and frequency scaling (DVFS) technology, enabling them to adjust their operating voltage and frequency in accordance with the current workload. This adaptive capability not only optimizes performance but also significantly boosts energy savings, making it especially beneficial for mobile devices that experience fluctuating processing demands. By balancing power and performance, ARM processors play a pivotal role in enhancing the longevity and usability of portable electronics.

8.2.3.3 Performance

Numerous ARM processors are designed to support multi-core configurations, allowing for parallel processing that significantly enhances performance in multitasking environments. A prime example is ARM's big.LITTLE architecture, which strategically pairs high-performance cores with energy-efficient cores. This combination optimizes processing based on the specific demands of different tasks, ensuring that power-consuming applications run efficiently while less demanding tasks utilize minimal resources.

In addition to this innovative architecture, ARM has incorporated advanced features such as the ARMv8-A architecture's support for 64-bit processing. This capability greatly improves performance for applications that require extensive memory addressing, such as databases and large-scale enterprise applications. With 64-bit processing, systems can handle more memory, enabling them to manage larger datasets and support more complex computational tasks. This synergy of multi-core configurations and advanced processing features makes ARM processors particularly adept at addressing the diverse needs of modern computing, from mobile devices to high-performance servers.

8.2.3.4 Applications

ARM processors are the backbone of most mobile phones (e.g., smartphones and tablets), IoT devices, embedded systems, edge computing, and data centers. To this end, they power popular smartphones like the Apple iPhone (using custom ARM-based A-series chips) and Samsung Galaxy devices (using Exynos and Qualcomm Snapdragon processors). ARM technology is prevalent in IoT applications due to its low power requirements. Devices like smart thermostats, wearables, and industrial sensors often use ARM Cortex-M microcontrollers for efficient data processing.

Many embedded systems, such as automotive controls, industrial automation, and consumer electronics, utilize ARM processors. The ability to integrate with various peripherals and support real-time operating systems makes ARM a go-to choice for these applications. ARM chips are increasingly used in edge computing devices to process data locally. For example, NVIDIA's Jetson platform (Mittal 2019), built on ARM architecture, is used for robotics and AI applications, leveraging its GPU capabilities for high-performance processing at the edge.

ARM's entry into the server market has gained momentum with companies like Amazon using ARM-based Graviton processors in their data centers. These processors offer a cost-effective and energy-efficient option for cloud services and scalable applications. In the context of supercomputers, a prime example is the Fugaku supercomputer in Japan, which, according to the Top500 list published in June 2024, is ranked as the fourth fastest supercomputer in the world and is entirely powered by Fujitsu's ARM-based A64FX processors.

8.2.3.5 Ecosystem and Development

ARM architecture supports a wide range of operating systems, including Android, Linux, and Windows, creating a robust ecosystem for software developers. Many development tools, libraries, and frameworks are optimized for ARM, making it easier to create applications. Additionally, the ARM ecosystem benefits from a large community of developers and extensive documentation, enabling rapid innovation and collaboration.

ARM-based technology has transformed the landscape of processing, particularly in mobile, IoT, and edge computing applications. Its energy-efficient design, scalability, and strong ecosystem have made it a preferred choice for developers and manufacturers looking to optimize performance while minimizing power demand and energy consumption. As technology advances, ARM's role is likely to grow even further, especially in areas requiring high efficiency and performance convergence such as cloud and high-performance computing data centers.

8.2.4 AI-Based EMS

AI-driven energy management systems (EMS) in data centers are essential for optimizing performance, lowering operational costs, and promoting sustainability. As discussed in previous chapters, data centers are major energy consumers, and the integration of AI technologies can greatly enhance their efficiency. For example, Google reported achieving an impressive average power usage effectiveness (PUE) of 1.1 across its global data centers, largely attributed to innovations in artificial intelligence and machine learning (Google data centers - efficiency 2024). In the following sections, a comprehensive overview of AI applications in this context will be provided.

8.2.4.1 Energy Consumption Monitoring

Modern data centers utilize thousands to millions of sensors and IoT devices that capture critical key performance indicators (KPIs) across a range of time scales—milliseconds, seconds, minutes, and hours. In this landscape, real-time data analytics powered by AI systems play a vital role in continuously monitoring energy consumption across various components, such as servers, cooling systems, and storage devices. This immediate analysis not only highlights areas of excessive energy use but also uncovers inefficiencies that can be addressed to optimize overall operations.

In addition to real-time monitoring, AI enables predictive maintenance by analyzing historical performance data. This capability allows for forecasting when equipment is likely to fail or require servicing. For instance, if a cooling unit shows a significant increase in energy usage compared to its historical data, AI can promptly notify the DCO of potential anomalies, allowing them to investigate and resolve issues proactively before they escalate into costly failures. This dual approach—real-time monitoring and predictive maintenance—not only enhances energy efficiency but also contributes to the overall reliability and longevity of data center operations.

8.2.4.2 Dynamic Cooling Management

When examining energy consumption in data centers, cooling systems emerge as the most significant contributors, sometimes accounting for up to half of the total energy usage required to maintain optimal temperatures for resources of information and communication technologies (ICT). In response to this challenge, adaptive cooling systems integrated with AI algorithms can effectively optimize cooling needs by analyzing environmental conditions such as temperature, humidity, and the specific workloads of servers. For instance, Microsoft's Natick project (Sutherland and Bopp 2023) showcases an innovative approach where AI manages the cooling processes in underwater data centers, dynamically adjusting based on real-time environmental data. This level of adaptability ensures that cooling efforts are both efficient and responsive to current conditions.

Moreover, AI can enhance airflow management in data centers by optimizing the layout of server racks and airflow patterns between hot and cold aisles. By continuously monitoring temperatures and fine-tuning cooling flows, these intelligent systems can significantly reduce energy waste. For example, if a particular server rack is generating excess heat, AI can adjust the cooling flow directed towards that rack while simultaneously reducing airflow to less active areas. This precise control not only boosts energy efficiency but also prolongs the lifespan of equipment, ultimately leading to a more sustainable and cost-effective data center operation.

For instance, by employing DeepMind's AI technology, Google achieved a 40% reduction in the energy used for cooling their data centers. The AI system analyzes data from thousands of sensors to optimize cooling in real-time. Another use case is eBay where the company implemented AI to optimize its data center cooling systems, resulting in a 20% reduction in energy consumption. The system uses machine learning to analyze temperature data and make real-time adjustments.

8.2.4.3 Workload Optimization

The ICT resources such as servers, storage systems, and networking equipment play a crucial role in managing and executing the workloads of data center clients. These workloads are the primary drivers behind a data center's energy consumption, as they underpin essential activities like computation, communication, and data storage.

To address the energy consumption associated with these workloads, intelligent workload distribution powered by AI can be employed. This technology enables real-time analysis of workloads, facilitating their efficient distribution across servers to enhance energy efficiency. For instance, when certain servers experience low utilization, AI can dynamically migrate workloads to these less active servers, optimizing energy consumption throughout the entire infrastructure.

A striking example of this innovation is Google's DeepMind, which has successfully integrated AI into its data center operations. By employing advanced workload management strategies and predictive analytics, DeepMind has achieved a remarkable 30% reduction in energy consumption. This not only demonstrates the potential for AI to transform data center efficiencies but also highlights a critical step toward sustainable energy practices in the IT industry. By adopting such intelligent systems, data centers can significantly mitigate their environmental impact while continuing to meet the growing demands of their IT customers.

8.2.4.4 Energy Efficiency Algorithms

By harnessing historical data alongside machine learning, AI algorithms can effectively forecast peak usage periods and proactively optimize resource allocation. This strategic foresight enables efficient deployment of energy resources, alleviating the strain on specific servers and significantly reducing waste. Furthermore, those algorithms can identify patterns in energy consumption and recommend changes by dynamically adjusting power usage based on workload and optimizing energy draw during off-peak times.

Reinforcement learning (RL) techniques can determine the optimal CPU frequency through dynamic voltage and frequency scaling (DVFS) technology, tailoring it to the specific demands of the workload. This approach not only enhances performance but also contributes to energy savings by ensuring that the CPU operates at the most efficient level based on real-time requirements.

Ultimately, by meticulously examining historical energy consumption data, AI can uncover valuable insights into the configurations that yield the highest energy efficiency. This analysis empowers DCOs to make well-informed decisions regarding hardware and infrastructure investments. For instance, AI can identify patterns that reveal which server setups or cooling systems minimize energy usage while maximizing performance. By leveraging AI-driven recommendations, operators can not only reduce operational costs but also enhance the sustainability of their facilities. This proactive approach ensures that investments are strategically aligned with long-term energy efficiency goals, ultimately contributing to a greener and more cost-effective data center environment.

8.2.4.5 Integration with RES and Automated Reporting

AI can help data centers integrate renewable energy sources (RES) by predicting energy generation from on-site solar panels or grid-wide windmills, allowing for more efficient use of these resources. For instance, companies providing services like Amazon's AWS are using AI to schedule workloads based on the availability of renewable energy. Furthermore, AI can optimize the charging and discharging of energy storage systems (like batteries) to align with renewable energy availability, ensuring that data centers use green energy whenever possible.

Regarding energy reporting, AI can automate the collection and reporting of energy consumption metrics, helping data centers comply with environmental regulations and sustainability goals. Additionally, AI can aid in calculating the carbon footprint of data center operations, enabling organizations to track their sustainability initiatives and make data-driven decisions to reduce emissions. This reduces the administrative burden and improves transparency.

8.2.4.6 Future Trends

As AI technology improves, more DCOs will adopt in the future AI-driven solutions for their energy management systems, leading to enhanced efficiency and reduced operational costs. Furthermore, the rise of edge computing (see Sect. 8.2.2) will necessitate more localized energy management solutions. AI will play a key role in optimizing energy use across distributed data centers. Additionally, as organizations prioritize sustainability, AI will help data centers achieve net-zero emissions by optimizing energy use and integrating more renewable sources.

AI-based energy management in data centers is and will continue to transform how energy consumption is monitored, optimized, and reported. By utilizing real-time analytics, predictive maintenance, dynamic cooling management, and intelligent workload distribution, data centers can reduce energy costs, minimize environmental impact, and enhance operational efficiency. As the demand for data centers continues to rise, AI will be pivotal in ensuring that they operate sustainably and efficiently.

8.3 Data Centers: The Paradox of Progress in Modern Society

As we reach the final section of this book, I would like to share some concluding thoughts and insights on its content, which primarily addresses the dual role of data centers in our contemporary society. They serve as powerful technological assets while simultaneously posing significant societal challenges. This complexity calls for a more thorough exploration, which various chapters of this book have addressed. In the following discussion, I will highlight and summarize several key aspects.

8.3.1 Economic Impact

Data centers are essential to the economy, underpinning a wide range of sectors by offering vital infrastructure for scientific computations, cloud services, e-commerce, and social media. They generate employment opportunities across various fields, including technology, construction, maintenance, and energy.

A prime example is Amazon's AWS, which employs thousands of people and has a substantial impact on the local economies surrounding its data centers. By fostering job creation and supporting local businesses, data centers are key players in driving economic growth and innovation.

8.3.2 Environmental Concerns

While data centers offer considerable economic advantages, their environmental impact is a growing concern. They are significant contributors to greenhouse gas emissions and face scrutiny for their substantial energy consumption. According to a report from the international energy agency (IEA), the electricity demand of data centers could surge by three to four times by 2030 if we fail to implement meaningful energy efficiency improvements.

This projection underscores the urgent need for innovative solutions, such as optimizing cooling systems, adopting advanced energy management technologies, and investing in renewable energy sources. As data centers expand to meet increasing digital demands, striking a balance between economic growth and environmental responsibility will be crucial for a sustainable future.

8.3.3 Sustainability Initiatives

In light of these environmental challenges, numerous technology giants are proactively launching sustainability initiatives. Companies such as Google and Microsoft have pledged to operate their data centers entirely on renewable energy. For instance, Google has been securing renewable energy through long-term contracts, allowing it to align its energy consumption directly with renewable sources. This strategic approach not only reduces their carbon footprint but also sets a precedent for the industry, showcasing a commitment to sustainability that other companies are increasingly motivated to follow.

8.3.4 Cooling Innovations

Innovative cooling solutions are emerging as a vital strategy to lower energy consumption in data centers. A noteworthy example is Meta's adoption of a technique known as "free cooling." This method leverages ambient outdoor air to cool servers, eliminating the need for conventional air conditioning systems. As a result, Meta has achieved substantial energy savings and significantly reduced its power usage effectiveness (PUE) values, a key metric for measuring energy efficiency in cloud computing data centers.

By integrating free cooling technologies, Meta not only minimizes its carbon footprint but also enhances the overall sustainability of its operations. This approach exemplifies how data centers can harness natural resources and advanced engineering to optimize performance while addressing environmental challenges. As more companies explore similar innovations, we can expect a transformative shift towards greener and more efficient data center operations in the industry.

8.3.5 AI and Efficiency

Artificial intelligence (AI) is instrumental in boosting the efficiency of data centers. Organizations are increasingly harnessing machine learning algorithms to better predict and manage their energy demands. By sifting through historical data, AI not only fine-tunes cooling systems to operate more efficiently but also anticipates potential hardware failures before they arise.

This proactive approach minimizes downtime and substantially lowers maintenance expenses. For example, some data centers are utilizing AI-driven analytics to dynamically adjust cooling based on real-time workload demands, ensuring optimal temperature regulation. Additionally, predictive maintenance powered by AI can alert technicians to potential issues, enabling timely interventions that prevent costly outages and enhance overall operational reliability.

8.3.6 Regulatory and Social Responsibilities

As data centers continue to expand, regulatory bodies are placing greater emphasis on their environmental footprint. There is a noticeable trend toward advocating for stricter energy efficiency standards and regulations aimed at reducing carbon emissions associated with these facilities. Furthermore, rising public awareness about sustainability issues is significantly shaping consumer preferences.

This growing consciousness is prompting companies to implement more environmentally friendly practices to meet consumer expectations. For instance, many tech giants are now committing to using renewable energy sources, investing in energy-efficient technologies, and even seeking certifications for their green initiatives, all in an effort to align with the values of their environmentally-conscious customers.

8.3.7 Future Orientations and Conclusion

Looking ahead, the trajectory points toward hyper-efficient data centers that leverage cutting-edge technologies such as edge computing. By processing data closer to its origin, edge computing significantly reduces both latency and energy consumption. This shift toward decentralized models—where data processing occurs at local edge nodes—has the potential to ease the burden on centralized data centers. Additionally, the development and deployment of intelligent energy management systems will be crucial. These systems aim not only to minimize energy usage but also to reduce peak demand and actively engage in demand-side management initiatives, the details of which were given and discussed in different chapters of this book.

In conclusion, while data centers are integral to our contemporary digital framework, they pose considerable environmental and energy challenges. The future will hinge on the industry's capacity to innovate and embrace sustainable practices, transforming data centers into more harmonious contributors to society rather than merely burdensome necessities. Achieving this balance will be essential to ensuring that data centers function as a positive force in our increasingly digital landscape. This book equipped both theorists and practitioners in the data center field with the necessary knowledge and scientific methodologies to optimize operations and navigate the complexities of this evolving domain.

References

Google data centers - efficiency. https://www.google.com/about/datacenters/efficiency/. Accessed 07 Nov 2024

Green500 list - june 2024. https://top500.org/lists/green500/list/2024/06/. Accessed 15 Oct 2024

Top500 list - June 2024. https://top500.org/lists/top500/list/2024/06/. Accessed 15 Oct 2024

Acun B, Lee B, Kazhamiaka F, Maeng K, Gupta U, Chakkaravarthy M, Brooks D, Wu C-J (2023) Carbon explorer: a holistic framework for designing carbon aware datacenters. In: Proceedings of the 28th ACM international conference on architectural support for programming languages and operating systems, vol 2, pp 118–132

Bae YH, Kim KO, Choi BH (2010) Lake sihwa tidal power plant project. Ocean Eng 37(5–6):454–463

Basmadjian R, Boubouh K, Boussetta A, Guerraoui R, Maurer A (2022) On the advantages of p2p ml on mobile devices. In: Proceedings of the thirteenth ACM international conference on future energy systems, pp 338–353

Bonsor K (2001) How hydropower plants work. Retrieved 31 Jan 2013

Boubouh K, Basmadjian R, Ardakanian O, Maurer A, Guerraoui R (2023a) Efficacy of temporal and spatial abstraction for training accurate machine learning models: a case study in smart thermostats. Energy Build 296:113377

Boubouh K, Basmadjian R, Ardakanian O, Maurer A, Guerraoui R (2023b) Efficient and accurate peer-to-peer training of machine learning based home thermal models. In: Proceedings of the 14th ACM international conference on future energy systems, pp 524–529

Boubouh K, Basmadjian R, Ardakanian O, Maurer A, Guerraoui R (2023c) Peptm: an efficient and accurate personalized p2p learning algorithm for home thermal modeling. Energies 16(18):6594

Jaye D (2024) A new era for the old dominion: strategies for the virginia state government to lead an equitable & ambitious energy transition. PhD thesis, Massachusetts Institute of Technology

Kong X, Wu Y, Wang H, Xia F (2022) Edge computing for internet of everything: a survey. IEEE Internet Things J 9(23):23472–23485

Liu Y, Peng M, Shou G, Chen Y, Chen S (2020) Toward edge intelligence: multiaccess edge computing for 5g and internet of things. IEEE Internet Things J 7(8):6722–6747

Mittal S (2019) A survey on optimized implementation of deep learning models on the nvidia jetson platform. J Syst Arch 97:428–442

Pierson J-M, Baudic G, Caux S, Celik B, Da Costa G, Grange L, Haddad M, Lecuivre J, Nicod J-M, Philippe L et al (2019) Datazero: datacenter with zero emission and robust management using renewable energy. IEEE Access 7:103209–103230

Sanyal SK, Enedy SL (2011) Fifty years of power generation at the geysers geothermal field, California–the lessons learned. In: 36th workshop on geothermal reservoir engineering, Stanford, California

Saul J (2024) Data centers face seven-year wait for dominion power hookups. Bloomberg

Sutherland T, Bopp G (2023) The pacific futures of subsea data centers. New Media & Soc 25(2):345–360

Temiz M, Dincer I (2022) A newly developed solar-based cogeneration system with energy storage and heat recovery for sustainable data centers: energy and exergy analyses. Sustain Energy Technol Assess 52:102145

U. I.-c. A. D. Note (2014) Hornsea offshore wind farm project one

Yu W, Liu Y, Dillon T, Rahayu W (2022) Edge computing-assisted iot framework with an autoencoder for fault detection in manufacturing predictive maintenance. IEEE Trans Ind Inf 19(4):5701–5710

Glossary

Agent Is an entity that acts autonomously to perform specific tasks or make decisions based on its environment or predefined rules. In technology, agents can refer to software programs or algorithms that operate on behalf of users or systems.

Anomaly Refers to something that deviates from what is standard, normal, or expected. In various contexts, it can indicate an unusual occurrence or a significant outlier that warrants further investigation.

Application programming interface Is a set of rules and protocols that allows different software applications to communicate with each other. It defines the methods and data formats that applications can use to request and exchange information.

Artificial intelligence Is the simulation of human intelligence processes by computer systems. It encompasses a range of technologies that enable machines to perform tasks that typically require human cognition, such as understanding natural language, recognizing patterns, solving problems, and making decisions.

Auction Is a public sale in which goods or property are sold to the highest bidder. Participants place competitive bids, and the item is sold to the person who offers the most money.

Automation framework Refers to a structured approach that utilizes tools, scripts, and processes to automate the management, monitoring, and orchestration of data center operations. This framework helps streamline tasks such as provisioning, scaling, configuration management, and monitoring, enhancing efficiency and reducing human error.

Bandwidth Refers to the maximum rate at which data can be transmitted over a network connection in a given amount of time, typically measured in bits per second (bps). It determines how much information can be sent or received simultaneously, influencing the speed and quality of internet connections, streaming, and downloads.

Big Data Refers to the vast volumes of structured and unstructured data that are too large or complex for traditional data processing tools to handle efficiently. It

R. Basmadjian, *Demand-Side Energy and Power Management in Data Centers*, Lecture Notes in Energy 1, https://doi.org/10.1007/978-3-031-85956-4

involves three key characteristics, often called the "three Vs" of volume, velocity, and variety.

Blueprint Is a detailed technical drawing or plan used primarily in architecture and engineering. It typically includes precise measurements, layouts, and specifications for constructing buildings or machines.

Bottleneck Is a point in a process where the flow of work is restricted, leading to delays and reduced overall efficiency. This limitation can occur in various contexts, such as manufacturing, computer networks, or project management.

Capacitor Is an electronic component that stores and releases electrical energy. It consists of two conductive plates separated by an insulating material called a dielectric. When voltage is applied, an electric field develops across the plates, allowing the capacitor to store charge.

Capital expenditure Refers to the funds a company invests in acquiring, upgrading, or maintaining physical assets such as property, buildings, or equipment. These are typically one-time expenses aimed at long-term benefits.

Carbon footprint Measures the total greenhouse gas emissions, particularly carbon dioxide (CO_2), directly and indirectly associated with an individual, organization, event, or product. It is typically expressed in terms of carbon dioxide equivalents (CO_2e) to account for different gases' global warming potential.

Cache Is a high-speed storage layer that temporarily holds frequently accessed data for quick retrieval, improving overall system performance. It acts as an intermediary between the CPU and main memory (RAM), allowing the processor to access data more rapidly than if it had to fetch it from slower memory.

Cloud computing Is the delivery of computing services-such as storage, processing power, and applications-over the internet (the "cloud"). Instead of relying on local servers or personal devices, users can access resources and services remotely, allowing for flexibility, scalability, and cost-effectiveness. For example, platforms like Google Drive or Dropbox enable users to store and share files online, while services like AWS (Amazon Web Services) and Microsoft Azure provide businesses with scalable infrastructure for hosting applications and data.

Conservative governor Similar to onDemand, however it increases the frequency more gradually, making it more suitable for battery-powered devices that require smoother scaling.

Contractual terms Are the specific provisions and conditions laid out in a contract that define the rights, obligations, and expectations of the parties involved. These terms ensure clarity and legal enforceability in agreements.

Control theory Is a mathematical framework that studies how systems can be influenced or regulated to achieve desired behaviors. It involves designing controllers that analyze the difference between a system's current state and its target state, then adjusting inputs to minimize this difference.

CPU core Is an individual processing unit within a CPU (central processing unit) that can execute instructions independently. Modern CPUs often contain multiple cores, allowing them to handle multiple tasks simultaneously through parallel processing.

CPU frequency Measured in hertz (Hz), indicates how many cycles per second a processor can execute, influencing its speed and performance. For example, a CPU with a frequency of 3.0 GHz can perform 3 billion cycles per second. Higher frequencies typically lead to faster processing speeds, allowing for quicker calculations and more tasks completed in a given timeframe. However, frequency alone doesn't determine overall performance, as factors like core count and architecture also play critical roles.

Cyber-physical system Integrates computational elements with physical processes, enabling real-time monitoring and control. As an example of a cyber-physical system, Smart Grid synchronizes electricity distribution with real-time energy consumption data, optimizing efficiency and reliability. In essence, such systems bridge the digital and physical worlds, enhancing functionality and responsiveness across various domains.

Cybersecurity Involves protecting computer systems, networks, and data from digital attacks, theft, or damage. It encompasses measures like firewalls, encryption, and intrusion detection systems to safeguard sensitive information. Strong cybersecurity is vital for maintaining privacy, trust, and operational integrity in today's digital landscape.

Dashboard Is a visual display of key metrics and data points that provides an at-a-glance view of performance or progress on specific objectives. It often consolidates information from various sources, making it easier to analyze trends, track KPIs (Key Performance Indicators), and make informed decisions.

Database Is an organized collection of structured information that allows for easy storage, retrieval, and management of data. It enables users to efficiently store vast amounts of information and perform complex queries.

Data bus Is a communication system that transfers data between components inside a computer or between computers. It consists of a set of physical connections (wires or traces) that carry data, allowing different parts of a system-like the CPU, memory, and peripherals-to communicate efficiently.

Data center A centralized facility containing numerous computer servers, storage systems, and networking equipment for processing, storing, and managing data. It features advanced infrastructure for power, cooling, security, and connectivity, supporting applications like cloud and high-performance computing, enterprise IT, and website hosting.

Data privacy Refers to the proper handling, processing, storage, and sharing of personal information. It involves protecting individuals' data from unauthorized access, misuse, or breaches. Laws like the GDPR in Europe aim to strengthen data privacy rights and give individuals more control over their information.

Demand-side management Refers to strategies and programs aimed at influencing consumer demand for energy. Its goal is to optimize energy use, reduce peak demand, and lower costs. DSM helps balance supply and demand while promoting sustainability.

Die of a processor Is the small, square or rectangular piece of silicon that houses the integrated circuits of a microprocessor. It's the physical substrate where the processor's electronic components-such as transistors, capacitors, and resistors-are fabricated.

Digital circuits Are electronic circuits that operate using discrete values, typically represented as binary (0s and 1s). Unlike analog circuits, which can process a continuous range of signals, digital circuits focus on precise values, making them more reliable for processing data.

Distributed systems Are collections of independent computers that work together to achieve a common goal. They share resources, communicate through a network, and appear to users as a single coherent system, even though the components may be spread across different locations.

Enclosure In a data center, enclosure refers to protective structures that house IT equipment, such as servers and networking gear. These enclosures are designed to optimize airflow and cooling, enhance security, and support organization. Additionally, some enclosures have built-in cooling systems or use thermal management technologies to maintain optimal operating temperatures, like hot aisle/cold aisle configurations.

Energy transition Refers to the global shift from fossil fuel-based energy sources to renewable and sustainable energy systems. The goal is to reduce carbon emissions, combat climate change, and create a more sustainable energy future.

Ethernet Is a widely used networking technology that facilitates communication between devices in a local area network (LAN). It operates primarily over twisted pair or fiber optic cables and uses a protocol that defines how data packets are formatted and transmitted. Ethernet supports various speeds, ranging from 10 Mbps in older versions to 100 Gbps and beyond in modern configurations.

Governor in Linux Is a mechanism that manages CPU frequency scaling to optimize performance and power demand. It dynamically adjusts the processor's clock speed based on current workload demands

Graphics processing unit Is a specialized processor designed to accelerate the rendering of images and video. While originally created to handle graphics tasks, modern GPUs are also used for parallel processing in various computational tasks, making them essential for gaming, video editing, and increasingly, machine learning.

Holistic view Is an approach that considers the whole system rather than just its individual parts. It emphasizes the interconnectedness and relationships within a system. For example, in environmental studies, it might involve examining how ecosystems, human activity, and climate change interact, rather than isolating each factor.

Hypervisor Also known as a virtual machine monitor (VMM), is software that creates and manages virtual machines (VMs) on a physical host. It allows multiple operating systems to run concurrently on a single hardware platform by abstracting the underlying resources, such as CPU, memory, and storage. Hypervisors play a crucial role in virtualization, enabling improved resource utilization and isolation of workloads.

Incentives Are rewards or benefits designed to motivate individuals or groups to take specific actions or achieve certain goals. They can be financial, such as bonuses or raises, or non-financial, like recognition programs or additional time off.

Internet Is a global network of interconnected computers that enables the exchange of information and communication across vast distances. It operates using standardized protocols, such as TCP/IP, allowing diverse devices, from smartphones to servers, to connect and share data. The Internet underpins countless applications, including social media, e-commerce, and online education, transforming how we interact, learn, and conduct business.

Interoperability Is the ability of different systems, devices, or software to communicate, exchange data, and work together effectively, regardless of their underlying technologies. This concept is crucial in areas like healthcare, or Smart Grid.

Kernel of an operating system Is the core component responsible for managing system resources and facilitating communication between hardware and software. It acts as an intermediary, allowing applications to interact with the hardware without needing to manage the hardware directly.

Latency Is the time delay between a user's action and the response from a system, often measured in milliseconds (ms). It reflects the time it takes for data to travel from the source to the destination and back, impacting the responsiveness of online activities.

Linux Is an open-source operating system (OS) based on the Unix architecture, known for its flexibility, robustness, and security. It was created by Linus Torvalds in 1991 and has since evolved into a widely used OS for servers, desktops, and embedded systems.

Logical operations Are fundamental operations in computer science and mathematics that manipulate boolean values-true or false (1 or 0). They are used in decision-making processes within programming, algorithms, and circuit design. The primary logical operations include AND, OR, XOR, and NOT.

Machine learning Is a subset of artificial intelligence that enables computers to learn from data and improve their performance on specific tasks without being explicitly programmed. By using algorithms and statistical models, machine learning systems can identify patterns, make predictions, and adapt over time based on new information.

Markov chain Is a specific type of stochastic process that features the Markov property. This memoryless characteristic makes Markov chains particularly useful for modeling systems that evolve randomly over time.

Markov property States that the future state of the stochastic process depends solely on its current state and not on the previous states.

Markovian analysis Is a mathematical approach used to model systems that undergo transitions from one state to another, having the Markov property.

Markovian process Is a stochastic process characterized by the Markov property, which states that the future state of the process depends only on its present state, not on the sequence of events that led to it. This memoryless property simplifies analysis and modeling.

Memoryless property Refers to a characteristic of certain stochastic processes, particularly in the context of probability distributions like the exponential distribution and the geometric distribution. It indicates that the future behavior of the process is independent of its past states.

Methodology Is a structured approach or set of principles used to guide the planning, execution, and evaluation of a project or process. It outlines the steps, techniques, and tools to be employed, ensuring consistency and effectiveness in achieving desired outcomes.

Motherboard Is a crucial component of a computer that serves as the central hub connecting all hardware parts. It houses the CPU (central processing unit), RAM (random access memory), and provides slots for expansion cards like GPUs (graphics processing units) and sound cards.

Network Is a collection of interconnected devices, such as computers, servers, and routers, that communicate and share resources. On a larger scale, the internet is a vast network of networks that connects millions of devices globally, enabling activities like browsing the web, streaming videos, etc. Networks can be wired, using Ethernet cables, or wireless, using Wi-Fi technology.

NP-hard Refers to a class of problems in computational complexity theory that are at least as hard as the hardest problems in NP (nondeterministic polynomial time). While NP problems can be verified quickly given a solution, NP-hard problems do not necessarily have a quick verification process.

Ondemand governor It scales the CPU frequency dynamically based on current load. It increases frequency when the load rises and decreases it when the load decreases, offering a balanced approach between performance and power savings.

Operating system Is software that acts as an intermediary between computer hardware and the user. It manages hardware resources, such as the CPU, memory, and storage, allowing applications to function effectively. The operating system provides a user interface, either graphical (GUI) or command-line (CLI), for interaction. Examples include Windows, macOS, and Linux. Key functions of an OS include process management, memory management, file system management, and device management, ensuring that different programs and users can operate smoothly and securely on the same system.

Operational expenditure Refers to the ongoing costs associated with running a business or organization on a day-to-day basis. These expenses include costs for labor, materials, utilities, and rent, among others.

Orchestrator Is a tool or system that automates and manages the coordination of different services, applications, or processes within a complex environment, often in cloud-based or microservices architectures. Its main role is to streamline workflows, ensuring that various components work together efficiently.

Packet Is a formatted unit of data carried by a packet-switched network. It includes several key components such as header, payload, and trailer. For instance, when you send an email, the email data is divided into packets, each labeled with the sender and receiver information, allowing them to be routed efficiently across the Internet.

Performance governor It sets the CPU to run at maximum frequency continuously, ensuring the highest performance but at the cost of increased power demand and energy consumption.

Poisson distribution Is a probability distribution that describes the number of events occurring within a fixed interval of time or space, given that these events happen with a known constant mean rate (λ) and independently of the time since the last event.

Poisson process Is a statistical model that describes events occurring randomly over a fixed period of time or space, where these events happen independently of each other. It is particularly useful for modeling situations where events are rare and happen at a constant average rate.

Port Is a physical connection point where devices connect to the switch to communicate within a network. Each port can connect to one device at a time, such as a computer, printer, or another switch.

Powersave governor In contrast to the Performance governor, this mode keeps the CPU at the lowest frequency, prioritizing energy efficiency over performance.

Predictive analytics Involves using statistical algorithms and machine learning techniques to analyze historical data and forecast future outcomes. By identifying patterns and trends, it enables organizations to make informed decisions and anticipate potential events.

Quantum computing Is a cutting-edge technology that leverages the principles of quantum mechanics to process information. Unlike classical computers that use bits (0s and 1s), quantum computers utilize qubits (quantum bits), which can exist in multiple states simultaneously due to phenomena like superposition and entanglement.

Quantum mechanics Is a fundamental theory in physics that describes the behavior of matter and energy at the smallest scales, such as atoms and subatomic particles. It challenges classical intuitions and introduces concepts that can seem counterintuitive.

Queue Refers to the arrangement of individuals or items waiting for service or to be processed in a specific order. It embodies the principle of First In, First Out (FIFO), where the first person to arrive is the first one to be served.

Regression Is a statistical method used to understand the relationship between a dependent variable and one or more independent variables. For example, a simple linear regression could explore how a person's weight (dependent variable) is influenced by their height (independent variable). By fitting a line to the data points, weight can be predicted based on height, helping to make informed decisions or forecasts. In more complex cases, like multiple regression, factors like age and activity level could be included to create a more accurate model.

Renewables Are energy types that are naturally replenished and can be harnessed without depleting natural resources. Examples include solar, wind, hydropower, biomass, geothermal, and tidal energy sources.

Round-robin Is a scheduling algorithm used primarily in time-sharing systems, where each process is assigned a fixed time slot in a cyclic order. This method ensures that all processes receive equal CPU time, making it fair and efficient for managing multiple tasks.

Semiconductors Are materials with electrical conductivity between that of conductors (like metals) and insulators (like glass). They are the foundation of electronic components such as transistors, diodes, and integrated circuits (ICs), which are used in everything from computers and smartphones to solar cells and LEDs. Their ability to control electrical signals makes them vital in the development of modern technology.

Server Is a powerful computer or system that provides data, resources, or services to other computers, known as clients, over a network. For example, a web server stores and delivers web pages to users' browsers when they request them. Another example is a file server, which allows multiple users to access and share files across a network.

Service level agreement Is a formal document that outlines the expectations and responsibilities between a service provider and a client. It specifies the quality, availability, and responsibilities of the service to ensure both parties have a clear understanding of what is to be delivered.

Smart grid An advanced electrical grid system that utilizes digital technology, communication networks, and data analytics to enhance the efficiency, reliability, and sustainability of electricity distribution. It integrates traditional energy sources with renewable energy, enabling real-time monitoring and management of energy flows.

Smart meter Is an advanced energy meter that records electricity, gas, or water usage in real-time and transmits this data to both consumers and utility companies using wireless communication. Unlike traditional meters, which require manual readings, smart meters provide timely information on energy consumption, enabling users to monitor their usage patterns and make informed decisions to save energy and reduce costs.

Statistical methods Are techniques used to collect, analyze, interpret, and present data. These methods help researchers and analysts make sense of data, identify patterns, and draw conclusions. There are different types of such methods like descriptive statistics (e.g., mean, median), inferential statistics, regression analysis (e.g., ARIMA), and analysis of variance.

Steady-state probabilities Refer to the long-term probabilities of a Markov chain being in each of its possible states after a large number of transitions. Once a Markov chain reaches its steady state, the probabilities of being in each state stabilize and no longer change over time, regardless of the initial distribution of states.

Stochastic process Is a mathematical object used to describe systems that evolve over time in a probabilistic manner. Unlike deterministic processes, where outcomes are predictable given initial conditions, stochastic processes incorporate randomness and uncertainty, making them suitable for modeling real-world phenomena.

Supercomputer Is an advanced high-performance computing data center designed to perform extremely complex calculations at incredibly high speeds. Typically used for tasks that require immense processing power, such as climate modeling, molecular simulations, and large-scale data analysis, supercomputers can perform trillions of calculations per second (measured in FLOPS - floating-point operations per second).

Sustainability Refers to the practice of meeting present needs without compromising the ability of future generations to meet their own needs. It encompasses environmental, social, and economic dimensions, focusing on responsible resource use, reducing waste, and promoting social equity.

Time-series Is a sequence of data points collected or recorded at specific time intervals, typically in chronological order. This type of data is used to track changes over time and identify trends, patterns, and seasonal variations.

Transistors Are semiconductor devices that act as electronic switches or amplifiers, controlling the flow of electrical current. They are fundamental building blocks of modern electronics, used in everything from computers to smartphones. Transistors work by using a small input voltage to control a larger output current, making them essential for processing and storing data in digital circuits.

Workload Refers to the amount of processing demand placed on the processing unit, typically quantified in terms of tasks, requests, or transactions that the CPU must handle over a specific period. It reflects the CPU's capacity to perform operations and the efficiency of resource utilization.

The manufacturer's authorised representative in the EU is Springer Nature Customer Service Centre GmbH, Europaplatz 3, 69115 Heidelberg, Germany. If you have any concerns regarding our products, please contact ProductSafety@springernature.com

Printed and bound by CPI Group (UK) Ltd, Croydon, CR0 4YY
30/09/2026
02225454-0002